The Open Future

The Open Future

Why Future Contingents are All False

PATRICK TODD

Great Clarendon Street, Oxford, OX2 6DP,
United Kingdom

Oxford University Press is a department of the University of Oxford.
It furthers the University's objective of excellence in research, scholarship,
and education by publishing worldwide. Oxford is a registered trade mark of
Oxford University Press in the UK and in certain other countries

First Edition published in 2021

Impression: 2

Published in the United States of America by Oxford University Press
198 Madison Avenue, New York, NY 10016, United States of America

British Library Cataloguing in Publication Data
Data available

Library of Congress Control Number: 2021931974

ISBN 978–0–19–289791–6

DOI: 10.1093/oso/9780192897916.001.0001

Printed and bound in Great Britain by
Clays Ltd, Elcograf S.p.A.

Preface

The problem of future contingents is a horrible, intractable problem, and I'm not sure what I was thinking when, a few years ago, I decided to try to write a book about it. And yet here we are. I hope I've managed to produce a book that is at least not wholly on the wrong track.

I began work on this book in the summer of 2018 as the Edinburgh visiting professor at Dartmouth College. However, my interest in the problem of future contingents began as an undergraduate, where I encountered the problem in connection to the problems of fatalism and free will. It struck me then—and continues to strike me now—that there could be nothing on the basis of which future contingents would be true, and so the best thing to say is that they just aren't true. I accepted in some semi-conscious way that this put me in conflict with the classical logic that I had just been taught—but I reasoned that other philosophers thought that this was a fine path to take, and anyway if it was good enough for Aristotle, it was probably good enough for me. Nevertheless, it struck me as vaguely ad hoc and unsatisfying to be forced to abandon certain *logical* principles in order preserve some quaint notion of "free will". If logic should imply that no one has free will, who are we simply to "revise" our logic? That was, if I recall, about as far as I got on this problem as an undergraduate.

As a graduate student, however, Kenny Boyce introduced me to A.N. Prior's "Peircean" view that future contingents are all *false*, and, in view of the difficulties just noted, I have been attracted to some version of an "all false" view ever since. Gradually, however, I became dissatisfied with what struck me as the stipulative character of Peirceanism, with its sheer insistence that *will* means "determined". Late in graduate school, and as a post-doc, I returned to the problem, and began developing a way of getting to an "all false" view, but not by courtesy of Peirceanism. The view I developed depended crucially on a comparison with Russell's theory of definite descriptions, and in 2013, as a new member of staff at the University of Edinburgh, *Mind* accepted my paper, "Future Contingents are All False! On Behalf of a Russellian Open Future", my first real foray into debates about the logic of future contingents.

Naturally, it all went downhill from there. For the acceptance of that paper coincided with my becoming friends with a group of new colleagues—all philosophers of language and logic—whose insights forced a fundamental reorientation of my approach to the problem of future contingents: Anders Schoubye (now at Stockholm), Bryan Pickel (now at Glasgow), and especially Brian Rabern (thankfully still at Edinburgh). The story of this book really begins here, as innumerable conversations with Anders, Bryan, and Brian turned what was otherwise, for me, a sort of side-interest in the logic of future contingents into a full-blown obsession, and forced in me a recognition that the theory, as developed thus far, was at the very least woefully under-defended. On a more personal level: it is not merely that, without their help, I wouldn't have been *able* to write a book of this kind—though that much is true. It is also that, without their help, I never would have seen fit to *try* to write a book of this kind.

But let me back up. There is much I still agree with in that *Mind* paper, but much that now strikes me as naïve or otherwise wrongheaded. (It has taken a good few years for me to become willing to write that sentence.) Concerning the *Mind* paper, one of my critics—Jacek Wawer—despite being severely critical of my "extremism", was kind enough to remark that "the elegant simplicity of this theory is impressive". The appearance of this book was significantly delayed by my emotional attachment to what I saw as this elegance—and my feeling that there just *had* to be a deep connection between debates about definite descriptions and debates about future contingents. In point of fact, however, the view which Wawer characterized as elegantly simple was not precisely the final view I had adopted in that paper. In fact, in order to preserve the truth of "future necessities", I had to make my final proposal fundamentally *disjunctive.* And this was never going to be ultimately satisfying.

Nevertheless, in the first draft of this book, I was still defending an updated version of the Russellian view I defended in the *Mind* paper, with one full chapter devoted to responding to the published objections from Schoubye and Rabern (2017) and Jacek Wawer (2018). In the end, and for reasons I won't bother elaborating, it took friendly suggestions from Kenny Boyce, Matt McGrath, Matthew Mandelkern, and (as ever) Brian Rabern finally to convince me that the spirit of my view is better captured, not by saying that *will* is, in part, some kind of disguised definite description, but instead is better captured by saying that *will* is a universal quantifier over what (in Chapter 2) I call the *available* futures—the causally possible futures consistent with the primitive future directed facts. This shift has resulted, I believe, in a number of crucial benefits to the book now before you; besides being intrinsically more plausible, it is (a) more in keeping with approaches to *will* found in linguistics, (b) fits better with my key proposal that *will* is a "neg-raiser", and (c) makes my core comparison with Lewis' semantics for the counterfactual much more straightforward. There is much in common between the view set out in this book (especially in Chapters 2–4) and the view developed in the *Mind* paper—indeed, at certain level of description, they are logically equivalent—but much that is crucially different, and, I hope, much that is crucially better, while retaining all the same simple elegance (such as it was) of the former theory.

Part of what makes the problem of future contingents so difficult, and so intractable, is that the problem touches on so many disparate areas of philosophy.

Indeed, the core device I appeal to in Chapter 3 ("neg-raising")—what is perhaps *the* key to my defense of my whole theory—has its home, not in philosophy at all, but in linguistics. I certainly do not claim that expertise in all of the needed areas has miraculously converged upon *me*. However far I've gotten in this book (and the reader must judge how far that is), it is because I've been guided, at crucial points, by the expertise of others.

I've also just been fortunate enough to interact with people likely to be able to help me. At lunch after a talk Daniel Rothschild gave in Edinburgh, conversation turned—or, if you must know, I turned the conversation—to what I was working on at the moment; in what seemed an offhand remark, Rothschild noted that what I was talking about sounded like "neg-raising"—a comment I made certain to follow up. And I soon discovered, to my great excitement, that there was already a *name* for the phenomenon I had previously been struggling so hard to identify. This then led to my paper, parts of which now form the basis of Chapter 3, "The Problem of Future Contingents: Scoping out a Solution", published in *Synthese*. Not only was I extremely fortunate to have Rothschild mention neg-raising, I was also extremely fortunate to have Laurence Horn—perhaps the chief authority on neg-raising—as a sympathetic (albeit nevertheless critical) referee. Horn says that Anselm is the "patron saint of neg-raising"—and I am certainly not going to contradict Horn on this matter. But even if Anselm is the patron saint of neg-raising, Horn certainly is its chief confessor, and I took a great deal of encouragement from our interactions. The neg-raising idea, even if ultimately mistaken, I thought, isn't *catastrophically* or *obviously* mistaken, and that idea, I felt, deserved a hearing.

I am deeply grateful to the University of Edinburgh (and to Mike Ridge) for a research support grant that allowed me to organize and host a workshop on my first draft of this book in June 2019, which led to very substantial improvements (some of which were noted above). In particular, for their participation in this workshop, I wish to thank Dilip Ninan, Derek Ball, Stephan Torre, Liv Coombes, Matthew McGrath, Mona Simion, Mark Thakkar, Stephan Torre, and my Edinburgh colleagues Brian Rabern and Wolfgang Schwarz. Concerning Wolfgang's comments in particular: I will be the first to admit that I cannot fully resolve what should be said about the interaction of *will* and *probably* (and the associated set of issues about credence discussed in Chapter 6). Conversations with Wolfgang, however, have encouraged me to think that there *is* a solution for the open futurist in this domain. Those conversations, however, have equally encouraged me to think that, if someone is going to fully say what that is, that person is not going to be me.

I presented an early draft of Chapter 4 at the Desert Philosophy Workshop (St. George, Utah) in April 2019; I wish to thank Josh Armstrong, Will Starr, Eliot Michaelson, Grace Helton, Shyam Nair, and James Shaw for especially helpful feedback. I also wish to thank audiences at departmental colloquia at Queens

University Belfast (April 2018), Dartmouth (July 2018), and Nottingham (November 2018). Special thanks are also very much due to Matthew Mandelkern, Richard Woodward, and Kenny Boyce, all of whom read the entire first draft and gave me enormously helpful and detailed comments on practically every chapter.

It is tempting to elaborate on the details concerning the way in which the following people have helped me over the years, but I shall have to content myself with the following: For helpful conversations and critical comments, I wish to thank, besides those already mentioned above, Fabrizio Cariani, Paolo Santorio, Geoff Pullum, Jon Gajewski, David Plunkett, Ghislain Guigon, Mark Balaguer, Neal Tognazzini, Philip Swenson, Andrew Bailey, Bradley Rettler, Jon Kvanvig, Sam Levey, David Hunt, Mike Rea, Sven Rosenkranz, Alex Pruss, Robbie Williams, Aldo Frigerio, Alda Mari, Jacek Wawer, Matt Benton, Godehard Brüntrup, Georg Gasser, Michael Nelson, John Perry, and John Martin Fischer.

I am extremely grateful to two anonymous referees for Oxford University Press, and can only apologize for the fact—and I'm afraid it is a fact—that I haven't always been able fully to do justice to their trenchant comments, especially concerning Chapter 1. In my defense, theirs are not the only comments I have failed to address; indeed, I have failed to address a whole range of comments it would probably have been sensible to address, but I simply didn't know how to address.

I am also deeply grateful to my colleague Brian Rabern for making the flowcharts in Chapters 2 and 4, and for allowing me to reprint our joint paper in *Noûs* as Chapter 7. In fact, I would be remiss if I didn't return to more explicitly acknowledge my enormous debt to Brian, whose patient advice, expertise, and good philosophical sense have shaped this book in more ways than I could mention. It would be impossible for me to catalogue all the ways in which some good point in this book ultimately traces back to some suggestion from Brian, or to catalogue all the errors that would almost certainly have appeared in this book, but for his influence. Indeed, it is only slightly an exaggeration to say that if a point is made in this book at all, it is either because Brian agrees with it, or couldn't convince me out of it. It goes without saying, of course, that any errors that *do* remain in this book are accordingly mostly his fault.

Finally, I am most grateful to my wife Rebecca, whose unfailing support was crucial to my completion of this project, and whose patience for delicate scope distinctions I have strained on numerous occasions. And before I forget, I must also thank my dog Yora and young son Gabriel, for doing their utmost to distract me from the task at hand.

Contents

The Open Future: Introduction to a Classical Approach

At least since Aristotle's famous 'sea-battle' passages in *On Interpretation 9*, some substantial minority of philosophers has been attracted to the thesis that future contingent propositions—roughly, propositions saying of causally undetermined events that they will happen—systematically fail to be true. However, open futurists, in this sense of the term, have always struggled to articulate how their view interacts with standard principles of classical logic—most notably, with bivalence and the Law of Excluded Middle. For consider the following two claims:

There will be a sea-battle tomorrow

There will not be a sea-battle tomorrow

According to the kind of open futurist at issue, both of these claims may currently fail to be true. In this sense, the future is "open". According to many, however, denying the disjunction of these claims ("There will be a sea-battle tomorrow or there will not be a sea-battle tomorrow") is tantamount to denying the Law of Excluded Middle (LEM). Accordingly, the open futurist must either deny LEM outright, or instead maintain that a disjunction can be true without either of its disjuncts being true. Worse, according to bivalence, if a proposition is not true, it is false—and thus the open futurist seemingly must either deny bivalence, or instead maintain that a disjunction can be true although both of its disjuncts are *false*.

Such are the familiar problems. The thesis of this book is that they are born of an illusion. The thesis of this book is that the disjunction of the above two claims is no instance of the Law of Excluded Middle—indeed, the thesis of this book is that the disjunction of the above two claims is not an instance of any principle whose validity is ultimately worth accepting. In this book, I do not defend a denial of LEM, and I do not defend the truth of the given disjunction by way of defending the claim that the disjunction is true even though its disjuncts fail to be true. I defend the claim that the disjunction is no instance of LEM, and that, in the relevant contexts, this disjunction is simply *false*, because both of its disjuncts are false. (With apologies to the reader, I do qualify this claim in Chapter 2 [*Missing Ambiguities*?]—but these qualifications can wait.) The central goal of this book is

The Open Future: Why Future Contingents are All False. Patrick Todd, Oxford University Press. © Patrick Todd 2021.
DOI: 10.1093/oso/9780192897916.003.0001

to defend the thesis that *future contingents are systematically false*. I thus defend a version of the doctrine of the open future that is *consistent* with the classical principles of bivalence and Excluded Middle.

The thesis that future contingents are systematically false has been defended before. Indeed, such a view was first put forward by Charles Hartshorne in 1941, and later defended by A.N. Prior in the 1950s and 1960s in the form of his 'Peircean' tense logic. My own view and the Peircean view thus have much in common: in particular, both maintain that future contingents are all false. However, in my estimation, the Peircean view is subject to serious objections. If Peirceanism were thus the only way of maintaining that future contingents are systematically false, then the open future *would* indeed require a revision of classical logic. My goal in this book is to articulate a version of the thesis that future contingents are all false that is not subject to the problems that plague the Peircean.

The central goal of this book is thus to develop a plausible, non-Peircean account of the open future and the semantics of future contingents that preserves classical logic. A brief word about this goal is in order. The standpoint that animates my discussion is not so much that classical logic is the true logic, but that the open future gives us no reason to think that it isn't. I am agnostic concerning whether there is any such thing as the "true" logic, and I am agnostic whether, if there is such a thing, classical logic is that logic. Nevertheless, it is clear that both bivalence and Excluded Middle still command the loyalty of a significant proportion of philosophers. I hope that the *desirability* of an account of the open future that preserves both such principles is not in need of substantial defense.

One distinctive feature of this book is the extent to which it features discussion of problems concerning future contingents and *omniscience*. It is, of course, common for theorists to point out the longstanding historical connections between the topics of future contingents and divine foreknowledge. As often as not, however—barring, of course, explicit discussion of this issue in the philosophy of religion—these connections are only noted in passing, and at any rate do no substantial work in motivating the relevant positions or arguments. As readers of this book will notice, however, considerations of divine omniscience are, starting in Chapter 6, invoked routinely in this book. This fact reflects my own interests in the philosophy of religion, but it also reflects my conviction that thinking about the problems of omniscience and the future are invaluable when assessing philosophical theories of the open future. Indeed, I believe that a primary advantage of the view I aim to develop is that it promises to provide an elegant story concerning omniscience and the future—and a primary argument I develop against rival ("supervaluationist" and "relativist") views is that they can tell no such story. However, a word of caution is in order about these points. At no point in this book do I develop arguments in which the *truth* of theism is invoked.

The result, I hope, is an essay that appeals to philosophers of religion, but is not itself a work directly in the philosophy of religion.

This book will assume the basic framework of A.N. Prior's tense logic. In particular, it will assume henceforth without comment that it is unproblematic to ask about the meaning, and the truth conditions, of propositions such as 'It was n units of time ago that *p*' and 'It will be in n units of time hence that *p*'—which, following Prior, I will abbreviate throughout as '**P**n*p*' and '**F**n*p*', respectively.

The book is organized around the resolution of what might be called "the problem of future contingents". Again, future contingent propositions are propositions saying of contingent, presently undetermined events that they will happen. (The events must be neither determined to occur, nor determined *not* to occur.) The problem of future contingents arises from the following conflict. On the one hand, we have what we might call *the grounding problem*. If nothing about present reality—and the laws governing how reality unfolds over time—settles it that the relevant events will happen, how and why is it *true* that they will happen? What, in short, accounts for the truth of future contingent propositions? Or if nothing *does* account for their truth, how are they nevertheless true? This is the grounding problem. On the other hand, we have what we might call *the logical problem* and a series of *practical problems*. If, instead, such propositions are never true, what becomes of the classical logical principles of bivalence and Excluded Middle? This is the logical problem. Further, if such propositions are never true—or even false—we seem to face a series of roughly *practical* problems regarding, for instance, our practices of betting, our credences regarding future contingents, our assertions about the future, and especially our practice of retrospectively predicating truth to predictions that in fact come to pass. If you predict that a horse will win a race, and then that horse does win, we will typically say that "you were right". If future contingents are never true, however, then it is not clear how this practice can make sense. These are our practical problems.

Open futurists endorse the grounding problem, and thus face the logical problem and the practical problems. In Chapter 1, I develop and advance the grounding problem. In Chapters 2–5, I address the logical problem for the open future. In Chapters 6–8, I respond to the practical problems. The result: the grounding problem stands, and the logical and practical problems can be addressed—and we have a defense of the doctrine of the open future.

It is worth noting that this book shall simply take for granted the two substantive theses that are plausibly necessary in order for the "grounding problem" to get off the ground: causal indeterminism, together with what might be called *no-futurism* in the ontology of time. In debates about the ontology of time, there are three primary competitors: *presentism*, the *growing-block* theory, and *eternalism*. Roughly speaking, presentism is the thesis that only present objects exist, the growing-block theory is the thesis that past and present (but no future) objects exist, and eternalism is the thesis that past, present, and future objects all exist. In this book, I assume non-eternalism. The argument of Chapter 1 is that, given either presentism or the growing-block theory, future contingents lack an appropriate sort of 'grounding', and therefore fail to be true. However, since my own personal view—which I shall at no point attempt to defend—is that *presentism* is true, I shall try in Chapter 1 to defend a *presentist* version of the open future that does not similarly result in an open past. Growing-blockers, however, can regard that project as a failure, and nevertheless accept the arguments to come for the claim that future contingents are all false.

This book also simply takes for granted the thesis of causal indeterminism. This is the thesis that the past and the present, together with the causal laws, fail to entail a unique future. That is, indeterminism is the thesis that there is at least more than one total way reality could evolve from "here", consistently with present reality and causal law. In keeping with tradition, we can call any total way things may go from a given moment that is consistent with causal law a *branch*. Thus, indeterminism is the thesis that there are multiple *branches*. An important word of caution, however: in this book, I am not thinking of these "branches" as in any way *concrete*. Rather, they are simply abstract representations—that is, abstract representations of total ways things could evolve. (In point of fact, they are segments of traditional abstract possible worlds.) Thus, to say that there are *branches* in my sense is not to commit oneself to the kind of *branching* at issue in some (so-called "many-worlds") interpretations of quantum mechanics.

Having now stated what this book simply assumes but does not defend, I am now in position to offer brief chapter by chapter summaries of what it does defend.

In Chapter 1, I develop what I above called the *grounding problem*, and articulate what I take to be the *metaphysical* case for the open future. More particularly, I argue that presentism and indeterminism imply the open future—or, in the terminology to come, I argue that, given presentism and indeterminism, there is no 'privileged branch' of those that remain causally possible. In this chapter, I investigate what sort of principles regarding *truth* and *grounding* together ought to imply that, given presentism and indeterminism, there are no truths regarding undetermined aspects of the future. I further respond to the problem that, given *presentism* and indeterminism, if we have an argument for the open future, we also have an unwelcome argument for the *open past*. It is worth noting that Chapter 1 is the only properly *metaphysical* chapter in this book. In Chapter 1, I attempt to argue on metaphysical grounds that there is no privileged future branch—but the rest of the book, by and large, simply takes for granted that there is no privileged future branch. Of course, if no arguments even in the *vicinity* of those of Chapter 1 are cogent arguments, what comes after Chapter 1 is perhaps of little interest. Nevertheless, the rest of the book can be read in isolation from Chapter 1.

In Chapter 2, I articulate *three models of the undetermined future.* In a context in which there are multiple future branches consistent with the past and the laws, is there any such thing as the "actual future"? According to the Ockhamist, there is an actual future history, and it is determinate which history is the actual future history. (Thus, there is a *privileged* future history.) According to the supervaluationist, there is an actual future history, but it is indeterminate which history is the actual future history. On the view I defend, however, there just *is no* "actual future history" in the first place. I further bring out the result that proponents of all three models can accept a plausible *modal* semantics for *will*—one on which *will* is a universal quantifier over all "available" branches. I show how this semantics for *will* combined with the various models under consideration gives rise to differing results about the truth-values for future contingents. In particular, I bring out the result that, if there are several *available* branches, then future contingents, given a plausible semantics for *will*, simply come out false.

In Chapter 3, I articulate my core response to the logical problem for open futurists. The central points I develop in Chapters 3–6 pertain, *inter alia*, to controversial distinctions in *scope.* The view I develop depends crucially on an important semantic distinction between the following two claims:

It is not the case that it will be in 20 minutes that there is rain. (~Fn*p*)

It will be in 20 minutes that there is no rain. (Fn~*p*)

My strategy is to defend the thesis that *will* is a so-called "neg-raising predicate". *I don't think that Trump is a good president* strongly tends to implicate *I think that Trump is not a good president*—although the former does not semantically entail the latter. The same goes, I believe, for *It is not the case that it will rain in 20 minutes* and *It will be in 20 minutes that there is no rain.* Under "standard" (viz., Ockhamist) assumptions about the future, the former would of course entail the latter—and it is for this reason, I contend, that we have such trouble hearing a distinction in meaning between the given claims. On metaphysical grounds, however, one might reject these standard assumptions. I further defend a series of scope distinctions that are predicted by a theory on which future contingents are all false.

In Chapter 4, I defend these scope distinctions (and my theory of the open future more generally) by means of a comparison with the counterfactual conditional. In Chapter 3, I attempt to make plausible a denial of the principle some authors have called "Will Excluded Middle" (WEM): Fn*p* ∨ Fn~*p.* As I hope to show, a denial of Will Excluded Middle is deeply parallel to the denial of what has been called "Conditional Excluded Middle":

(CEM) If it had been the case that *p*, it would have been the case that *q* ∨ If it had been the case that *p*, it would have been the case that ~*q*.

CEM has been the subject of vigorous dispute in both metaphysics and semantics. My claim is simple: if CEM is not a *semantic* truth—and many (e.g., Lewis and Williamson) have contended that it is not—then neither is WEM.

Chapter 5 takes up what A.N. Prior has called "The Formalities of Omniscience". The view I defend can accept the following biconditional: *p* if and only if *God believes p*. Accordingly, my view can happily accept the following biconditionals:

It was n units of time ago that *p* iff God (quasi-) remembers than n units of time ago, *p*. (**P**n*p* iff **Rem**n*p*)

It will be in n units of time hence that *p* iff God anticipates that in n units of time hence, *p*. (**F**n*p* iff **Ant**n*p*)

Accordingly, my view can happily accept—whereas other open future views cannot—the plausible thesis that *the logic of the tenses is the logic of perfect memory and anticipation*. I further discuss a set of principles regarding divine omniscience that are crucially related to the scope distinctions defended in Chapters 2–4. In particular, I defend the following slogan:

> For an omniscient being: Absence of memory implies memory of absence, but absence of anticipation does not imply anticipation of absence. (~**Rem**n*p* implies **Rem**n~*p*, but ~**Ant**n*p* does not imply **Ant**n~*p*)

For example: from the fact that God does not remember a sea-battle yesterday, we can, given the relevant idealizing assumptions about God, conclude that God remembers there being no sea-battle yesterday—and thus that there was no sea-battle yesterday. However, from the fact that God does not *anticipate* a sea-battle *tomorrow*, we cannot conclude that God anticipates the *absence* of a sea-battle tomorrow; God may have no anticipation as of yet either way. This is, on my view, precisely the asymmetry of openness between past and future.

In Chapter 6, I shift gears. Chapters 3–5 are primarily concerned with the logical problem for the open future. In Chapter 6, however, I turn to the first of our *practical* problems for the doctrine of the open future—a problem articulated in the first instance by A.N. Prior. Prior noted that it can *seem* that, on the open future view, if I bet that a given horse will win a race, and then that horse does win, someone working under open-futurist assumptions could refuse to grant the payout. After all, it would seem, what I was betting was true was not, on my view, true. I respond to this problem by developing a picture of *betting* that does not presuppose the truth of any future contingents—and I address a difficult related problem having to do with what *credence* we should assign to the claim that it will rain tomorrow, on assumption that no rain tomorrow is objectively possible. (Hint: it is 0.)

In Chapter 7, I build on themes from Chapter 5, and criticize the two most prominent rival positions to my own: *supervaluationism* and *relativism*. In this chapter, co-written with Brian Rabern (and previously published), I begin consideration of what might be called the "prediction problem". This problem is associated with a critical principle of tense-logic, a principle I call "Retro-closure": $p \rightarrow \mathbf{PF}p$—or, in its metric formulation, $p \rightarrow \mathbf{P}\mathrm{n}\mathbf{F}\mathrm{n}p$. More simply: If p, then it was n units of time ago that it will be the case n units of time later that p. For example: if it is raining, it follows that yesterday it would rain a day later. The crucial result from this chapter is the following. You can take your pick between the open future and Retro-closure, but—contra what is predicted by both supervaluationism and relativism—you *can't have both*. I further add an appendix to this chapter—written solely by myself—explaining why, as I see it, we don't *need* Retro-closure. The argument against the open future from the validity of the Retro-closure principle is, in my estimation, far and away the most common argument given against the doctrine of the open future. It is thus crucial for a successful defense of the open future that we see how Retro-closure may plausibly be denied.

In Chapter 8, I address what has sometimes been called the *assertion problem* for open future views. Roughly, the problem stems from the observation that what are plausibly future contingents are still sometimes properly *assertible*—despite being, on my view, false. The challenge is thus to specify how the open futurist's proposal interacts with standard norms of assertion. There are, to be sure, further objections to the doctrine of the open future—my version included—but a book has to end somewhere, and mine ends here.

1
Grounding the Open Future

My goal in this chapter is to provide an argument that, given presentism and indeterminism, the future is "open", although the past is not. One of the primary rejoinders to presentist arguments for the open future is that such arguments would *also* imply the open past—and that we are rightly reluctant to accept the open past. This much is right: we should not accept the open past. My aim in this chapter is to show that it is plausible on presentist indeterminist grounds to accept the open future, but not the open past. There is nothing approaching a decisive argument that presentism and indeterminism together imply the open future—let alone such an argument that *also* does not imply the open past. Nevertheless, I believe that this position remains substantially theoretically motivated, and here I wish to bring to light these motivations.

Clearly, what makes it *prima facie* difficult to give a *presentist* argument for the open future that does not imply the open past is simple: presentism is, in itself, symmetric with respect to past and future; it says that neither past nor future objects exist. If the openness of the future were meant to follow from something like the *non-existence* of the relevant class of objects, then presentism is going to imply an open past if it implies an open future. In this light, we can observe that an argument that the future is open whereas the past is not is certainly easiest to develop if we assume, not presentism, but instead the growing-block theory of time.[1] For then the relevant asymmetry is simply built into the ontology of the theory; as we'll see, we could then say that truths about the past supervene on reality, and we could deny that the truth of future contingents would similarly supervene on reality, and so we could deny that there are such truths. Consequently, the bulk of this book could happily proceed from the assumptions of indeterminism and the growing-block theory of time. However, I do not wish to proceed from the assumption of the growing-block theory of time, and this for two reasons. First, whether presentists can give a cogent argument for the open future that does not imply the open past is of intrinsic metaphysical interest. Second, I am a presentist. And whereas I regard the doctrine of the open future as a reasonable doctrine, I am much less inclined to view the doctrine of the open past as a reasonable doctrine. If any argument from presentist assumptions to the open future similarly implied that the *past* is open, I would regard this as a deep

[1] Cf. Diekemper 2005 for an extended development of this theme.

The Open Future: Why Future Contingents are All False. Patrick Todd, Oxford University Press. © Patrick Todd 2021.
DOI: 10.1093/oso/9780192897916.003.0002

problem for the doctrine of the open future. The current project concerns whether the presentist open futurist can overcome this deep problem. I believe that she can.

In debates about indeterminism and the open future, it is routine for theorists to employ the terminology of *branches*. In the context of causal indeterminism, I shall assume, we have various abstract *branches* that represent maximal ways things might go from a given point in time, consistently with what has happened up to that point and the causal laws. And a central question in debates about the status of *future contingents* is whether, in such a context, there is any such thing as the *privileged* branch—that is, of all the branches, the branch that uniquely (and determinately) has the status of being *going to obtain*—lit up with what Belnap and Green (1994) once called the 'thin red line'. My claim in this chapter is that there is *no privileged branch*, and therefore future contingents fail to be true, and the future is open.[2] Here I confront the question: how might we argue that there is not a privileged *future*, but there *is* a privileged *past*?

The central distinction I wish to develop in this chapter pertains to whether we ought to argue for the open future from a claim about *truth in general* or instead from a claim about *the future in particular.* I maintain that if we argue from a claim about *truth in general*—for instance, from the thesis that "truth supervenes on being" (TSB)—then, if we are presentists, we can accept the open future but *not* the open past only if we have (or are prepared to adopt) what many will regard as an exotic ontology. It is this result that I wish to bring out in the first section of this chapter. The spirit of my discussion is not to recommend such ontologies (nor is it to say anything *against* them)—rather, it is to show that, once we see how such ontologies would do the work TSB requires of them, we'll see that such work needn't be done in the first place. As I aim to argue, we should simply *reject* TSB, and instead maintain that the past is simply *brute*. However, our reasons for thinking that the past is brute (and TSB is false) are *not* reasons for thinking that the future is—or could be—similarly brute. We thus have substantial theoretical reason from presentist indeterminist assumptions to affirm the open future, but *not* the open past.

1.1 Truth and Reality

The most familiar arguments for the openness of the future proceed from some claim about the relationship between truth and reality—for example, from the

[2] It is worth noting that there are other conceptions of what it is for the future to be "open" than the one at issue in this book. For more on these themes, see Torre 2011 and Grandjean 2019. Further note: in this book, I am not thinking of "openness" as a sort of substantive pre-theoretical desideratum that various theories might be trying to capture, as in Barnes and Cameron 2009. For me, the sense of "openness" employed here is simply stipulative.

thesis that *truths require truthmakers.*[3] However, in light of well-known problems with "truthmaker maximalism"—the claim that all truths require truthmakers—it is worth seeing if we can produce an argument for the open future that assumes the *weakest* of such principles about truth: the thesis that *truth supervenes on being.*[4]

> TSB: Necessarily, for any true proposition, if that proposition were instead not true, there would have to be a difference in reality (in which objects exists or in what properties they instantiate).

It is of particular interest for our discussion how *presentism* interacts with TSB. If presentism is true, then *being* just is *present* being, and reality just is *present* reality; accordingly, if truth supervenes on being, it supervenes on how things are *right now.* Thus, according to presentism and TSB, we have:

> TSB-P: Necessarily, for any true proposition, if that proposition were instead not true, there would have to be a difference in present reality (in which objects presently exist, or in what properties they presently instantiate).

According to TSB, there can be no difference in truth without a difference in reality. But if all of reality is *present* reality, then TSB implies that there can be no difference in truth without a difference in present reality.

We can now give the following simple argument that, given TSB-P, future contingents cannot be true. Suppose for reductio that it were a true future contingent that there will be a sea-battle tomorrow. Since this is a true future *contingent,* it is not *determined* that there will be a sea-battle tomorrow (nor, of course, that there will not be a sea-battle tomorrow). But to say that it is not determined that there will be a sea-battle tomorrow is to say that what exists presently is consistent both with there being a sea-battle tomorrow and there *not* being a sea-battle tomorrow. Accordingly, if it is a true future contingent that

[3] Cf. Sider:

> More importantly, grounding the tenses in the present plus the laws of nature threatens to imply that the past is 'open', just as some have claimed that the future is open. If the laws of nature are present-to-past indeterministic, current facts plus the laws do not imply all the facts about the past; given presentism and either the truth-maker principle or the principle that truth supervenes on being, for many statements, φ, neither $\ulcorner$ it was the case that $\varphi \urcorner$ nor $\ulcorner$ it was the case that not-$\varphi \urcorner$ will be true. (Sider 2001: 38)

Markosian 1995, however, develops an argument for the open future from a theory of truth as *correspondence.*

[4] The literature on truth and truthmaking (and the related thesis that truth supervenes on being) is, of course, enormous. For a start on truthmaking (and a defense of truthmaker maximalism), see Armstrong 2004; for a critical discussion of these principles, see Merricks 2007. Bigelow states the 'truth supervenes on being' principle as follows: "If something is true then it would not be possible for it to be false unless either certain things were to exist which don't, or else certain things had not existed which do" (1988: 133).

there will be a sea-battle tomorrow, if this proposition were instead false, this would *not* require *any difference at all* in what presently exists and how things are. The present could be *just as it is*, and yet it be *false* that there will be a sea-battle tomorrow. Accordingly, if there are true future contingents, we have a violation of TSB-P. Given TSB-P, therefore, there can be no true future contingents.

Consider the familiar picture of the Laplacian demon. Suppose that, all of a sudden, a Laplacian demon comes into existence. And this Laplacian demon has comprehensive knowledge of (i) the state of the universe at the current moment and (ii) the laws of nature. And now its project is to construct a complete story of the future—or at least, to construct as much of such a story it is possible to construct from what is possible to gather about it from the present and the laws. Indeterminism is the thesis that any such Laplacian demon will not be able to recover a complete "story of the future" from current conditions and laws alone. Instead, that demon will be left with multiple such "stories" concerning which—again, looking solely at current reality and law—it has no reason to prefer one to the other. The consequence is this. If one such story is privileged over the others—if it is true that *that* story is the one that is going to obtain—then the fact that this story is the privileged one does not supervene on anything to which our demon has access. On presentist assumptions, however, this is to say that this truth does not supervene on *anything at all.*

Perhaps, however, taking a cue from Bigelow, there are ways of building something into the present from which a unique story of the future could in principle be constructed—crucially, something that does not violate the spirit of causal indeterminism. Perhaps, some presentists may say, things have (or some thing has) primitive *future directed properties*, properties like *being going to be in a sea-battle tomorrow*.[5] And if the relevant demon had access to *everything*, our demon would have access to such properties, properties from which it could predict a unique future. (The question whether causal determinism is true, these philosophers will insist, is thus a matter of whether the demon can predict a unique future from a more *limited* set of facts in the present, together with causal law—e.g., from the facts in the present, minus any facts about which things have which primitive future directed properties.) But now the relevant open futurists will be apt to complain. For presumably, no amount of inspection of a given naval commander is going to reveal his having a primitive property such as *being going to be involved in a sea-battle tomorrow*. To be sure, the demon may know that the commander has this property—and the commander may have this property. However, if he does so, they will insist, he will do so *because* his having it is entailed by *other* more basic facts about him, his causal environment, and the

[5] For a defense of this "Lucretian" way of grounding truths about the future and past, see Bigelow 1996, and Tallant and Ingram 2020. For a different presentist approach to grounding truths about the past, see Crisp 2007, and Ingram 2019.

causal laws. Accordingly, the open futurist can, with considerable justification, regard the postulation of such primitive future directed properties to be an ontological extravagance motivated solely by the preservation of a theory.

It is thus very plausible to suppose that, if indeed there are true future contingents, then the truth of these future contingents does not supervene on present reality, at any rate given natural assumptions about "present reality": present reality could be *just as it is*, and yet these propositions fail to be true. Accordingly, we may give our first argument for the open future:

1. If there are true future contingents, the truth of these future contingents would not supervene on present reality.
2. But all truth supervenes on present reality. So,
3. There are no true future contingents.

1.2 The Problem of the Past

But now our problems begin. As I see it, premise (1) of the above argument is plausible, and I do not here further consider ways one might dispute it.[6] The trouble, however, comes from the *bluntness* of the weapon employed in premise (2): TSB-P. For TSB-P requires that all truths supervene on present reality. What, then, becomes of truths about the *past*—in particular, truths about the past that presumably *cannot* be retrodicted from the present together with causal law? In other words: what about the truth of *past* contingents? This objection was well put by Michael Clark in the 1960s against Charles Hartshorne's defense of the open future. Clark writes:

> If the analysis is plausible for the future tense, why not a similar analysis for the past tense? If we consider present conditions as causal conditions, it is true that it is reasonable to regard the future as indeterminate in the sense that not every detail has been causally settled yet. So, given present states of affairs and the laws of nature, there are many different detailed courses which the future might take. But equally, given the laws of nature, if we consider present conditions as effects, there are many different detailed pasts from which they might have arisen. There are past events which are contingent with respect to the present, which are not

[6] Let me emphasize that I have only considered *one* candidate presentist way of attempting to show that future contingents might still "supervene on reality"—the Lucretrian strategy suggested by Bigelow. Perhaps the "ersatzist" strategy of Crisp (2007) might do better than Lucretianism in this regard. Briefly, however, my main contention, once more, is that the "ersatzist" strategy will not give the presentist open-futurist non-open paster (!) what he or she wants. If such a theorist said that there are no true future contingents because there are no relevant "ersatz future times", it seems clear that, on presentist grounds, neither will there be the relevant ersatz *past* times—and so, once again, if we're getting the open future, we are *also* getting the open past.

> retrodictable even in principle from present conditions using laws of nature. As time moves on the events at a given moment in the future are made more definite but those at a given moment in the past become less definite. Consider all the cases of suicide in Britain this year. After the suicides have occurred there remains evidence of them; the year after, the evidence is enough to determine precisely who killed himself, but hundreds of years later, perhaps, the remaining evidence is enough to establish roughly how many suicides there were but not precisely who they were. The openness of the future is matched by a corresponding openness in the past. (1969: 178)

I want to begin by considering how *certain* open futurists with certain ontologies *could* respond to the problem at issue. In the cited paper, Clark gives us a great many cogent objections to Hartshorne's account of future contingents. As it happens, however, this is an objection for which Hartshorne in particular was well prepared. Following A.N. Whitehead, Hartshorne was prepared to say that a complete, comprehensive story of the past *can* be retrodicted from the present. Did Hartshorne defend some temporal asymmetry thesis about the laws of nature? No. He maintained, with Whitehead, that the past can be retrodicted from the present independently of any such laws, namely from *God's memories*.[7] Thus, according to Hartshorne, though there are many *futures* consistent with present conditions and causal law, there is only one *past* consistent with present conditions and causal law (because there is only one such past consistent with present conditions). In this sense, Hartshorne simply denied that there were any true past contingents, and thus he escapes the charge that though he allows true past contingents, he does not allow true *future* contingents: he allows *neither*. If we (very much) strain our analogy, so long as we suppose that the Laplacian demon has access to the current state of God's memorial seemings, then that demon *could* retrodict a complete story of the past from those seemings. For given the relevant idealizing assumptions about God (assumptions, incidentally, Hartshorne already would have independently accepted), if we consider God's memories as effects, there is in fact only *one* detailed past from which those effects may have arisen. And since God is both eternal and incorruptible, so is the past—just as desired.

Of course, we might now ask the following. If there is only one complete past that is consistent with God's current memories, then why shouldn't we similarly say that there is only one complete *future* consistent with God's current *anticipations*? Why is it that God now has a complete set of memories that discriminate between all the possible pasts, but God has not yet formed a complete set of

[7] Whitehead 1929 (esp. pp. 12 and 347); Hartshorne 1970. Both Whitehead's and Hartshorne's views on this matter are complicated. For a helpful articulation of these aspects of "process theism", see Viney 2018, who maintains that Whitehead had a doctrine of the "objective immortality of the world in God"; see also Viney 1989: 84.

anticipations that discriminate between all the possible futures? But here Hartshorne might plausibly maintain that the asymmetry is grounded in the nature and the direction of *causation*. God has the relevant set of memories because the relevant events *caused* God to form those memories. But, he might insist, future events, in general, cannot *cause* one to form anticipations of those events—and this, in part, explains why God's anticipations are not comprehensive. For if God's anticipations are not effects of future events themselves, God is presumably forming such anticipations by means of *deductions* from current conditions. In other words, God is acting as something like the Laplacian demon imagined above, who predicts the future precisely via deduction from the present. Thus, if current conditions and laws do not fix a unique future (and *ex hypothesi*, they do not) then neither will God's anticipations pick out a unique future. The events God's memories are memories of *caused* God to form those memories, whereas the events God's anticipations are anticipations of do not *cause* God to form those anticipations. Hence the asymmetry. This, Hartshorne may say, is simply a fundamental feature of the direction of time.

Is this a satisfactory answer to the problem of the open past that arises if one defends the TSB-P-inspired argument for the open future given above? Well, on its own terms, perhaps it is. We start from an intuition that Truth Supervenes on Being. And since we are presentists, we maintain that this must be *present* Being. We then notice that, if there were truths about contingent aspects of the future, these truths would not supervene on being; and so we reason that there are no such truths. It is then objected: but presumably there are truths about contingent aspects of the *past*, and these truths would *also* not supervene on being. We respond: in the sense at stake, there are no truths about contingent aspects of the past, for the past is not *contingent* in the sense at stake: all of its details are fully retrodictable from current conditions, viz., from God's memories, and there are therefore *not* multiple pasts consistent with current conditions and laws. However, although God currently has a comprehensive set of memories specifying a unique past, God has no corresponding set of anticipations from which one could read off a unique future—and this asymmetry is principled, grounded in facts about the nature and direction of time, causation, or both.

A problem arises, however, when we notice that this theory seems to inherit the problems that *all* theories that appeal to the intentional states of an ideal being or beings to explain why there are truths of a certain kind seem to inherit. For suppose we asked the proponent of the theistic account above the following: what if God erased a memory—and, indeed, erased the memory of which memory he erased?[8] Such a procedure should be possible, we might have thought, for a being as powerful as *God*. One answer that has been given to this question recently

[8] Parallel question for the divine command theorist: what if God commanded something terrible?

(Rhoda 2009) is the one you might expect: God cannot erase a memory; for since God is perfectly rational, God never acts without sufficient reason, and necessarily so. But *God* could never have sufficient reason to erase a memory, and if not, it is impossible that such memories should be erased. Again, TSB has been preserved.

My point thus far is simple. Certain presentists *could* insist on TSB-P, and plausibly maintain the open future, but not the open past. But perhaps you are thinking what I'm thinking. For what this position seems to be bringing to light is that, even if we *could* preserve TSB in this way, TSB *needn't be* preserved in this way—or in any way at all. For notice where we end up on this position. There are truths about the past *only because* of God's current memories—and, in turn, *only because* God cannot (and certainly nothing else can) erase those memories. But is it really plausible to suppose that there are only truths about the past because there exist the relevant memories *now*? My feeling is that this is not plausible. My feeling is this. Perhaps Hartshorne is right. Perhaps God exists, and perhaps indeed God has the given memories, and perhaps God can't erase those memories. That would certainly be very interesting. Still. It is not *because* God exists with those memories that there are truths about the past. Per impossible, *should* God erase a memory, it is not like the truths about the past would go with them. But what we say about even God and God's memories we should say about *anything* in the present that is a causal trace of the past: truths about the past may in fact be verified by such traces—that would certainly be interesting if they were—but truths about the past *needn't be* verified by such traces. Those traces—whatever they are posited to be—could be erased, and yet the facts about the past would not be erased with them. It is, thus, the prerogative of truths about the past to be *brute* with respect to what exists in the present. But if what exists in the present is what exists simpliciter, it is the prerogative of truths about the past to be brute simpliciter.

The resulting picture is the following. If we were already presentist theists, then we should continue maintaining that truth supervenes on being. After all, theism provides a sort of trivial guarantee that truth supervenes on being; if for any truth, God believes that truth, and necessarily so, then if any such truth were instead not to be true, this would require a difference in reality, viz., a difference in God's beliefs. But we shouldn't say that there are truths about the past *because* these truths supervene on being, even if they do.

And if we were *not* already presentist theists, then we should simply reject Truth Supervenes on Being. We should deny Truth Supervenes on Being for the following reason. Suppose we agree that there was a sea-battle in 2019. But now consider the following counterfactual:

> (SBP) If it is true that there was a sea-battle in 2019, it would *still* be true that there was a sea-battle in 2019, even if everything went out of existence, and there came to be nothing at all.

I have the strong intuition that SBP expresses a truth. More generally, I have the strong intuition that the truths about the past would *remain* truths about the past, even if there came to be nothing at all. But if SBP expresses a truth, and presentism is true, then truth does not supervene on being. If it could come to be that nothing exists, you could have two worlds exactly alike in respect of what exists—viz., nothing—and yet in one world, it is true that there had been a sea-battle, and in the other, not true there had been a sea-battle. So truth does not supervene on being.

Let me clarify. When I say that, if it is true that there was a sea-battle in 2019, it would *still* be true that there had been a sea-battle in 2019, even if everything went out existence, I don't mean *everything*. In particular, I don't mean that if the claim that there was a sea-battle in 2019 went out of existence, the claim that there was a sea-battle in 2019 would still be true. What I mean is that if *everything on which the truth of this proposition could plausibly be thought to supervene* went out of existence, the claim would still be true. And this seems to me to be very plausible.

Consequently, if we are presentists, we should not attempt to argue for the open future and the non-open past by means of the general thesis that truth supervenes on being. For on one of the options considered above, we have found reason for presentists to think that TSB is simply false: truths about the past do not supervene on being, and yet there are such truths. And on *either* option we have explored above, we have said that, even if truths about the past *did* supervene on being, it is not *because* truths about the past supervene on being that there are truths about the past. But if it is not because truths about the past supervene on being that there are truths about the past, then it cannot be *because* truths about the future would *not* supervene on being that there cannot be truths about the future. Presentist arguments for an open future but a non-open past that proceed from a *general thesis about truth* seem doomed to fail. But then how *should* such presentists make their case, if they can make it at all?

1.3 The Future in Particular

I begin with the following observation. Philosophical arguments to the effect that it would be unacceptably *arbitrary* if there were truths of a certain kind are legion. And yet such arguments rarely seem to proceed from a general claim about truth. Let me give an example.

Nearly everyone agrees that there are no facts about what goes on in fictions beyond those entailed by a certain set of *other* facts—facts, for instance, having to do with authorial intent. Call such facts (whatever they are) the *fiction-determining facts*. In general, nearly everyone thinks that there are fiction-determining facts, and no facts about what goes on in fictions beyond those specified by the fiction-determining facts. We are reading the Harry Potter books. Harry sits down for breakfast at Hogwarts, though, naturally, the episode has little to do with breakfast; it is in fact merely an occasion for Harry to learn about the latest news from home. Nevertheless, it is mentioned (or implied) that on the table is orange juice, breakfast cereal, and much else. Harry hears the news as he has breakfast. Scene over. But now we wonder. Exactly what sort of breakfast did Harry have? The details weren't very much specified. What was the pulp content of the orange juice on the table? And what was its temperature as it sat there—exactly room temperature, higher, or lower? Was the milk skim milk, 1 percent, 2 percent, or whole? We have questions, and we want answers.

Nearly everyone agrees that, in this case, no such answers are to be had, and this is because there are no such answers in principle to be given. Unless the temperature of Harry's orange juice were somehow specified in the story, or perhaps entailed by facts specified in the story—then there is just *no fact of the matter* concerning the temperature of that orange juice.

But imagine encountering someone who believes that there *is* a fact of the matter concerning that temperature. To be sure, this person admits, we can never *know* what precise temperature that was, but this is not say that there is nothing there to know in the first place; there *is* something to know there, but we just don't know it. Accordingly, this person maintains, *God*—if there were a God—would know the answer to this question. Since God is *omniscient*, God knows all there is to know—and since there is something to know about the temperature of Harry's orange juice, God knows what that temperature was.

We are, of course, incredulous. Surely not even *God* knows what temperature Harry's orange juice was in the episode at issue—and surely that's not a failure of omniscience, but instead because *there is no fact of the matter* about this temperature. In other words, that's because "the temperature of Harry's orange juice during..." is an improper, denotation-less definite description that simply fails to refer. (More on this theme in the next chapter.) However, as we begin to express our incredulity, our theorist interrupts:

> You are hereby objecting to my claim that there is a truth (albeit one unknown to us) about what temperature Harry's orange juice was in the given episode at issue. But, in so objecting, you rely on a principle about *truth*, namely that truths require truthmakers; your claim—let me finish!—is that if there were a truth about the given temperature of the orange juice, that truth would lack a truthmaker. But the claim that all truths require truthmakers is disputable. What about negative existentials? You can't give me suitable truthmakers for *those*. And what about 'totality facts'? Reflection on such cases indicates that we should deny the claim that all truths require truthmakers.[9] Or perhaps your thought is

[9] "Negative existentials"—such as the claim that there are no hobbits—have proved difficult for "truthmaker maximalists". For discussion (esp. in the context of debates about presentism and truthmaking), see Asay and Baron 2014: 318–326.

> merely that *truth supervenes on being*—and that my truth would somehow fail to "supervene on being". However, as Trenton Merricks has argued, that principle is no better—or, as he may say, just as bad[10]—as the principle that truths require truthmakers. Consequently, you have no principled objection to my claim that there is some truth—unknown to us—concerning the temperature of Harry Potter's orange juice in the episode in question.

There is, of course, something absurd about this preemptive reply to our incredulity. Our inability to provide truthmakers for negative existentials is neither here nor there vis-à-vis our objection to this kind of fictional (hyper-) realism. For this is *not* how or why we are objecting to the claim that there is a fact of the matter in the given domain. That is, we are not arguing that

1. If there were truths about what goes on in fictions that are not specified by (. . .), then those truths would lack truthmakers/fail to supervene on being.
2. But all truths require truthmakers/supervene on being. So,
3. There are no truths about what goes on in fictions beyond those specified by (. . .).

If this *were* how we were arguing, then of course objections to premise (2) might be relevant—but it isn't, and they aren't. When we are objecting to this sort of hyper-realism, we are not proceeding from a claim about *truth in general*, but instead from a claim about *fictions in particular*.

If this *isn't* how are arguing, then how indeed are we arguing? What is it, then, about fictions in particular? Well, it is difficult to say. But we might begin, at least, by expressing intuitions like the following. We feel like facts about fictions are *derivative* in some way—that they *come from somewhere*. Again, we feel like facts about fictions are not *fundamental* facts, and accordingly, insofar as there are such facts, there are such facts insofar as there *other* facts that fix them or determine them.

I'll cut to the chase. When I am having the intuition that there are no facts about the future beyond those determined by the present, I am *not* having an intuition about truth in general. I am, rather, having an intuition about the future in particular—an intuition similar to the one nearly all of us already have about fictions. I am having the intuition that facts about the future stand to facts about the present and the laws as facts about fictions stand to the fiction-determining facts. That is, I have the feeling that the present and the laws *produce* the future—and I have the feeling that, since the present and the laws are all that there is, and since the present and the laws produce the future, there shouldn't be any facts about the future *beyond* those the present and the laws produce. That there should

[10] Merricks 2007: 68–97.

be such facts about the future—for instance, that there should *just be* a fact specifying that there will indeed be a sea-battle tomorrow, when everything about the present and the laws is as equally consistent with there being *no* sea-battle tomorrow—strikes me as mysterious and bizarre. For where would such a fact come from? Not from the future objects and events themselves; on presentist grounds, there just *are no* such objects and events. And not from facts about what the present and laws require. From where then? From nowhere. And so I am puzzled.

Perhaps this is not much of an argument. Indeed, I am wary of even calling it an argument. I am instead simply telling you that I think that facts about the future are derivative in some way from some more basic or fundamental facts, and giving you some indication of what I think those more basic or fundamental facts are, and hoping you agree. But if you do not think that facts about the future are or must be derivative in this way, then I am not sure what I could say to convince you. And that is fine.

But let me now return to the problem of the past. From the above, it should be evident that, if this is how we argue for the open future, we simply have no problem concerning why we do not similarly encounter an open past. For when I am having the intuition that the future is open, I am having the intuition that since the future is produced by, or explained by, or otherwise dependent on how things are *right now*, there are no facts about it beyond those entailed by how things are right now. However, I have *no feeling at all*—and, I suspect, hardly anyone has any feeling at all—that the *past* is produced by, or explained by, or otherwise dependent on how things are right now. Indeed, I have precisely the *opposite* feeling: my feeling instead is that the past is not at all produced or explained or dependent on the present. Indeed, as I argued above, I have the feeling that the present could be eliminated entirely, and yet the past facts would still be the past facts. But I have no inclination whatever to suppose that the present could be eliminated entirely, and yet the future facts still be the future facts.

Let me explain. Earlier we noted the plausibility of the following counterfactual to argue against the claim that truth supervenes on being:

> (SBP) If it is true that there was a sea-battle in 2019, it would *still* be true that there was a sea-battle in 2019, even if everything went out of existence, and there came to be nothing at all.

Accordingly, it is plausible to suppose that truths about the past are *brute* with respect to the present. But we now we must contrast the plausibility of SBP with the implausibility of its temporal mirror:

> (SBF) If it is true that there will be a sea-battle in 2219, it would *still* be true that there will be a sea-battle in 2219, even if everything went out of existence, and there came to be nothing at all.

SBF is, of course, patently ridiculous. If everything went out existence (in the next few minutes, say), it is not like it would *still* be true that there will be a sea-battle in 2219. If everything went out of existence in the next few minutes, it is hard to see how there could come to be a "2219", let alone a 2219 with a sea-battle. The lesson is this. The plausibility of SBP shows that it is the *prerogative of the truths about the past to be independent of what exists in the present.* Truths about the past, so to speak, just *don't care* what happens in and to the present. They are, therefore, brute with respect to the present—and, once more, if the present is all that there is, they are brute simpliciter. But our reasons for thinking that truths about the past are brute in this way are *not* reasons for thinking that truths about the future could be similarly brute. For whereas SBP is plausible, SBF is straightforwardly false.

1.4 Conclusion

We can sum up as follows. The substantial advantage of arguing for the open future by means of TSB is that TSB has a good claim on being a metaphysically neutral starting point. Indeed, many philosophers who seem to show no interest in arguing for the open future nevertheless are very much attracted to TSB. If we are presentist open futurists, however, TSB simply won't give us what we want. But our inability to argue for the open future from TSB does not leave us with no way of arguing for the open future, any more than it leaves us unable to argue against fictional hyper-realism. The trouble, of course, is that now we have little in the way of a metaphysically neutral starting point from which to proceed; we must simply motivate or otherwise make plausible our conception of the facts of the future being dependent on facts about the present and the laws. And this is what I aim to do in what follows.

2
Three Models of the Undetermined Future

In this chapter, I articulate three competing models of the undetermined future. The first model is roughly what is known as "Ockhamism", the second is "supervaluationism", and the third doesn't have a recognized name—but it is the model I defend. However, an important qualification: my aim in articulating these models is not primarily exegetical. If these models do not correspond to what you consider to be "Ockhamism" or "supervaluationism", that is fine. (However, if these models are not even remotely in the ballpark of what you consider to be "Ockhamism" or "supervaluationism", then at least one of us is seriously misinformed.) The central way in which these models differ concerns whether the models contain what we might call a *privileged branch* and what we might call an *actual branch*. The first has both, the second has only the latter, and the third has neither. In the end, I contend that the third model is to be preferred: we can make do without a privileged future history, and, indeed, without even an *actual* future history.

My other central aim in this chapter is to introduce and motivate a certain crucial semantics for *will*—a semantics on which *will* is a universal quantifier over all *available* branches. The question concerning the truth-values of future contingents thus becomes a (broadly) *metaphysical* question, the question of how many branches are genuinely "available". My contention, building on the argument of the previous chapter, is that there are no *primitive future directed facts*, and thus the available branches (in a context) just are the causally possible branches (in that context). The result is that future contingents are uniformly *false*.

2.1 Three Models

In this book, I simply take for granted the truth of causal indeterminism. Thus, all three models of the future I shall discuss take for granted that there are multiple causally possible histories (which, again, I shall also sometimes call "branches" or "futures") consistent with the past and the laws. These three models can be articulated as follows:

The Open Future: Why Future Contingents are All False. Patrick Todd, Oxford University Press. © Patrick Todd 2021.
DOI: 10.1093/oso/9780192897916.003.0003

There are multiple future histories consistent with the past and the laws, and

(I) exactly one of those histories is the actual future history, and it is determinate which history is the actual future history. (There is an actual future history, and it is determinate which.)
(II) exactly one of those histories is the actual future history, but it is indeterminate, for each given history, whether it is the actual future history. (There is an actual future history, but it is indeterminate which.)
(III) no given history is such that it is the actual future history. (There is no "actual future history" in the first place.)

Note that we can put the distinctions between these three models in different ways; e.g., in terms of *reference*. For instance, we can say that there are multiple future histories consistent with the past and the laws, and "the actual future history"

(I) refers to one such history, and it is determinate to which
(II) refers to one such history, but it is indeterminate to which
(III) does not refer

Similarly, and more concretely, suppose we stipulate that there is going to a be an indeterministic lottery tomorrow with three and only three possible winners: A, B, and C. We can then say that "the winner of tomorrow's lottery" either now

(I) refers to A, refers to B, or refers to C, and it is determinate to whom it refers
(II) refers to A, refers to B, or refers to C, but it is indeterminate to whom it refers
(III) does not refer.

At this stage, it is crucial to appreciate what models (I) and (II) have in common—and in turn what models (II) and (III) have in common. Now, what models (I) and (II) have in common is obvious. On both such models, *there is a unique actual future.* On model (I), it is determinate which future this is, and on model (II), it is indeterminate. And yet on both such models, *there is* such a future. Now, what models (II) and (III) have in common is slightly subtler. Model (I) maintains, whereas models (II) and (III) deny, what might be called the doctrine of *privilege.*

On model (I), although there are multiple future histories consistent with the past and the laws, it is determinate which history is the actual future history. Thus, this history has a certain privileged status over the others. However, on model (II), although there is a unique actual future history, it is *indeterminate* which history has this status. More particularly, every candidate future history is such that it is *indeterminate* whether it is the actual future history. Thus, in this respect, every such history is *on a par*.[1] On model (I), as we go through each candidate future history and ask, "Is this the unique actual future history?", at some point (although presumably we don't know where) the determinately correct answer is "Yes", and for all other such histories the answer is "No". However, on model (II), as we go through each candidate history and ask, "Is this the unique actual future history?", in each case *the answer is the same*—the answer is "That's indeterminate." Finally, on model (III), every future history consistent with the past and the laws is once more on a par. As we go through each such history and ask the relevant question, our answer is the same: "No". There is *no* unique actual future history, and so in each case, in answer to the question whether *this* is the relevant history, the answer has to be that it isn't. There just is no "actual future history" in the first place (although, perhaps, it will be that there is such a history!).

More particularly, suppose we stipulate that we have three and only three branches. In that case, we have the following respective answers to our question:

(I) Yes/No/No | No/Yes/No | No/No/Yes (these are the only three options, and one obtains, although perhaps we don't know which)
(II) Indeterminate/Indeterminate/Indeterminate
(III) No/No/No

Only on model (I) is some branch *privileged*. On the other two models, there is perfect parity amongst the branches. Models (I) and (II) are united in terms of there being a unique actual course of history; models (II) and (III) are united in terms of there being no *privileged* history in the model. Because models (II) and (III) share this feature, I shall therefore call them models on which the future is *open*. More particularly, we can define *the open future* as the denial of the doctrine of privilege. In this book, I defend model (III).

2.2 Future Directed Facts

I wish to construe the debate between models (I), (II), and (III) as in the first instance *metaphysical*—as a debate about the nature and existence of *primitive*

[1] Cf. Barnes and Cameron (2009: 296):

> Finally, it's determinately the case that exactly one of the worlds in {Futures} is actualized. One and only one world matches the complete atemporal state of the actual (concrete) world. It's just that it's indeterminate which of the worlds in {Futures} is in fact actualized. So at t it's still the case that determinately, there's a single, unique way the world atemporally is; it's just that it's indeterminate which (among a list of options) way is the way that the world atemporally is. That is, it's always determinately the case that the world is some particular way atemporally; it's just that (at least prior to the last moment in time) there's no single atemporal way that the world determinately is.

Cf. also Barnes and Cameron 2011: 2–3.

future directed facts—that is, facts specifying what is to come, which are *primitive* solely in the sense that they are not grounded in facts about current conditions and laws. As a first approximation, the dispute between models (I) and (II) is a dispute concerning the *nature* of these facts, whereas the dispute between models (I) and (II) and model (III) is a dispute concerning their *existence*.

The dispute between the Ockhamist and the supervaluationist is a dispute about whether the future directed facts are determinate or instead indeterminate. In other words, both parties agree that (a) there are future directed facts, and that (b) only one given future—the unique actual future—is consistent with the future directed facts. The dispute concerns whether it is determinate which future is uniquely consistent with the future directed facts, or instead indeterminate. In other words, the dispute concerns whether the future directed facts themselves are determinate, or instead indeterminate. If they are determinate, then only one given history is consistent with those facts, and it is determinate which history this is. If they are instead indeterminate, then whereas only one given history is consistent with those facts, it is indeterminate which history that is. The core idea shared by both models, however, is that there is something such that if you *add it* to the past and the laws, then a unique actual future is selected. (And these are—or are what I propose to call—the primitive future directed facts.) The difference, again, is simply whether it is determinate which future is selected, or instead indeterminate—although, again, on both such models, some unique future is selected.

The dispute between models (I) and (II) and model (III), however, is not a dispute concerning whether the primitive future directed facts are determinate or instead indeterminate; the dispute concerns whether there are any primitive future directed facts in the first place. On models (I) and (II), there are such future directed facts; on model (III), there just *are no* such facts.

In other words, on model (III), the histories consistent with the past and the laws and the future directed facts *just are*, well, the histories consistent with the past and the laws. On model (III), all reality has at its disposal, so to speak, to narrow down the class of possible worlds to the "actual world" is what has happened up to now, and the laws of nature. That's it. Thus: if what has happened up to now and the laws do not by themselves "narrow down" the class of possible worlds to one single world (that is, if indeterminism is true), then, on model (III), there just *is no* "actual world" at all. Reality does not, on this model, have the tools to give us one. On models (I) and (II), however, there is indeed a unique "actual world". Thus, reality does indeed have at its disposal something further, in addition to the past and the laws, to narrow down the class of possible worlds to a unique "actual world". And again, these are—or are what I propose to call—the (primitive) future directed facts.[2] (I sometimes drop the "primitive" qualifier

[2] Note: as I am conceiving of them, the role played by the primitive future directed facts with respect to *will* is analogous to the role some have seen for "counterfacts" with respect to the counterfactual *would*. For more on "counterfacts", see esp. Chapter 4.

hereon for simplicity.) On model (I), the future directed facts reality adds to the past and the laws are determinate, and so it is determinate which future is the unique actual future; on model (II), they are indeterminate, and so it is indeterminate which future is the unique actual future.

Our options can thus be re-stated as follows:

(I) There are many histories consistent with the past and the laws, but there is only one history consistent with the past and the laws *and the future directed facts*, and it is determinate which history this is.

Once you add in the future directed facts, there is only one history still available; and it is determinate which history this is.

(II) There are many histories consistent with the past and the laws, but there is only one history consistent with the past and the laws *and the future directed facts*, but it is indeterminate which history this is.

Once you add in the future directed facts, there is only one history still available; but it is indeterminate which history this is.

(III) There are many histories consistent with the past and the laws, and there are many histories consistent with the past and the laws *and the future directed facts*, because there *are no* future directed facts.

Once you add in the future directed facts, well, precisely nothing in addition is narrowed down—for there just are no future directed facts to "add in", as it were. Thus: on model (III), we might say, *the available futures just are the causally possible futures*. Indeed, it is here we can properly introduce the notion of an available future. The *available* futures I shall define as

> those futures that are consistent with the past and the laws and the future directed facts.

Thus, in this framework, according to model (I), there is only one available future, and it is determinate what it is, and according to model (II), there is only one available future, but it is indeterminate what it is, and according to model (III) there are exactly as many available futures as there are futures consistent with the past and the laws.

But here we must be careful. We must be careful not to employ this talk of "availability" as a rhetorical bludgeon against proponents of models (I) and (II). On models (I) and (II), yes, there is only one future still *available* (in the just defined sense of "available"). But this, we must insist, cannot at all be taken to commit proponents of such models to any form of *determinism*—and nor can we fairly accuse such models of violating the spirit of indeterminism, on grounds that, on this view, there is really only one future that is still "available". More particularly, given the relevant definition of "available", we cannot adequately characterize indeterminism as the thesis that there are, as of now, at least several *available* futures; whether there are several *available* futures is a separate matter entirely. Consider. Whether causal determinism obtains is solely a matter of whether there is or is not more than one future history consistent with the past and the laws of nature (whatever they are). And on models (I) and (II), the answer to this question is unambiguously "yes". Whether there is or is not more than one future history consistent with the past and the laws is, broadly speaking, a matter for the physicist. However, whether in a context in which there are many futures consistent with the past and the laws, there are *also* primitive future directed facts (which select for a unique actual future), is a matter for the metaphysician (and perhaps, as we shall shortly see, the philosopher of language). The important point here is the following: to say that on models (I) and (II), there is only one future that is "available" (viz., the unique actual future) is not at all to commit proponents of these models to the truth of causal determinism. On these models, many futures are causally possible.[3]

Two further notes—this time about "future directed facts". As I have described them, on models (I) and (II), there are primitive future directed facts, which facts are consistent with only one given causally possible future, viz., the unique actual future. However, it is important to note that I am not hereby committing the proponent of such models to any claim to the effect that these facts *ground* or *explain* the existence of a unique actual future. Indeed, proponents of these two models are perfectly free to maintain that, if anything, the order of explanation is quite the reverse: it is because there exists a unique actual future that there are (now) primitive future directed facts. In other words, consider the following:

(A) It is because there now exist the primitive future directed facts that there exists a unique actual future.

(B) It is because there exists a unique actual future that there now exist primitive future directed facts.

Models (I) and (II) are, in themselves, neutral on (A) or (B). Either way, however, on these models, we still have it that there is something such that, once it is "added to" to the past and the laws, it is consistent with only one of the various causally possible futures. Thus again: in this sense, on these models, only one future is "available".

[3] For more on this theme—that positing an actual future, in itself, is not definitionally inconsistent with indeterminism—see Borghini and Torrengo 2013.

A further note. Talking in terms of these "future directed facts" is perhaps unfamiliar and strange. But it is, I hope, not *too* strange. Suppose we asked the following (seemingly simple) question:

> Are there multiple ways things could be tomorrow consistently with all the facts about how things will be tomorrow?

This is, of course, an odd question to ask, but once we've asked it, we can see the following. On model (I), the answer to this question is "No". For the Ockhamist, given indeterminism, there are perhaps multiple ways things could be tomorrow, consistently with the past and the laws—but even then, there certainly are *not* multiple ways things could be tomorrow, consistently with all the facts about how things will be tomorrow. Again, once you add in all the facts about how things will be tomorrow, there is only one way things could be tomorrow. So similarly on model (II). However, on this view, since the facts about how things will be tomorrow are *indeterminate*, it is indeterminate which unique way things could be tomorrow is indeed *the* way things will be tomorrow. (There is indeed such a thing as *the unique total way things will be tomorrow*—it is just indeterminate to what we refer when we talk about the unique total way things will be tomorrow.) Again, however, the answer has to be "No". Finally, on model (III), the answer to this question is unambiguously "Yes". For model (III), there just are no facts about how things will be tomorrow that are not also facts about what the past and the laws require for tomorrow. Thus: "all the facts about how things will be tomorrow" simply amounts to "what the past and the laws require for tomorrow". But on the assumption of indeterminism, there are multiple ways things could be tomorrow, consistently with what the past and the laws require. Accordingly, there are multiple ways things could be tomorrow, consistently with all the facts about how things will be tomorrow.

2.3 The Thin Red Line

In their 1994 paper, "Indeterminism and the Thin Red Line", Belnap and Green introduced an influential bit of terminology into the literature on future contingents: the "thin red line". And so here we must ask: how is the "thin red line" related to the models described above? The unfortunate answer is: it's complicated. For Belnap and Green did not carefully distinguish the question of whether a certain branch is *privileged* and the question whether a certain branch is *actual.* Herewith how Belnap and Green introduced the notion of "the thin red line":

> One common way of conjoining the aspect of indeterminism formulated above with an understanding of our world deriving from a construal of the B-order as a

> series, is to hold that at a given moment in the history of our world from which there are a variety of ways in which affairs might carry on, one of those ways is asymmetrically privileged over against all others as being what is actually going to happen. Not only is it true that the coin could either come up heads or tails, and therefore true that it will be the case that either the coin comes up heads or the coin comes up tails. Not only is it the case that either the coin will come up heads, or the coin will come up tails. What is furthermore true is that there is, at the time of the toss, a directly referential, rigid, absolute specification of "what, at the time of the toss, is actually going to happen." This specification breaks the symmetry, picking out either Heads or Tails. Only our limited minds keep us from knowing which.
>
> It used to be said of the British Empire that it was maintained by a thin red line of soldiers in service to the Queen. We shall express the view just sketched by saying that from among the lines along which history might go, subsequent to an indeterministic moment, one of those lines is the course along which history will go, and it is both thin and red.
>
> . . . Further, belief in an actual future, a Thin Red Line, appears to be consistent with believing that there is indeterminism in the world.
>
> . . . We shall call the view that in spite of indeterminism, one neither needs nor can use a Thin Red Line, the doctrine of the *open future*. (1994: 366–367)

Now the point. On the one hand, the idea of the thin red line seems to be introduced as a way of saying that one history is "asymmetrically privileged over against all others". Thus, it might seem that to believe that there is a thin red line *just is* to believe in a privileged branch. On the other hand, the doctrine of the thin red line is *directly defined* in terms of there being an *actual future*: "belief in an actual future, a Thin Red Line".

The result is the following. To say that there is a "thin red line" is importantly ambiguous between models (I) and (II). If to say that there is a "thin red line" is to say that there is a *privileged* future history, then model (I) contains a thin red line, whereas models (II) and (III) do not. On the other hand, if to say that there is a "thin red line" is merely to say that there is an *actual* future history, then both models (I) and (II) contain a thin red line, whereas only model (III) does not. Recall, on model (II), though there exists a unique actual course of history, it is indeterminate which history has this status. Thus, though there is an *actual* history in the model, there is nevertheless a perfectly good sense in which there is no *privileged* history in the model. If to say that there is a thin red line is *simply* to say that there is an actual course of history (an actual future), then we would have to express model (II) as follows:

> There is a thin red line, but it is indeterminate which branch is the thin red line (or is "lit up with the thin red line").

Now, I have no objections to talking in these terms; at the same time, however, some might, and certainly I do not purport to own this terminology. That is, some might feel like the terminology was introduced precisely in order to introduce the notion of a *privileged* history—and so our statement above violates the spirit of that terminology. Finally, many authors, following Belnap and Green, define *the open future* as the denial of the doctrine of the thin red line. But such a definition leaves it unclear whether model (II) is a model on which the future is "open" (in the relevant sense). (Recall that I have defined it as the denial of the doctrine of *privilege*—and thus that model (II) counts as one on which the future is "open".) My proposal thus is simple: to refrain from employing the terminology of the thin red line.

Although in this book I will, for these reasons, bypass (for the most part) talk of the "thin red line", it is crucial to note the following. As I see it, when discussing these issues, many authors simply have not had model (II) in mind. (Indeed, in earlier writings on these topics, I was only faintly aware of it myself.[4]) The result is that, in my judgment, much of the literature systematically conflates the issue of whether there is a *privileged* branch with the issue of whether there is an *actual* branch. In other words, much (though certainly not all) of the literature tends to proceed without an explicit recognition of model (II). If models (I) and (III) are our only models, then there being a *privileged* branch and there being an *actual* branch come to the same thing: model (I) has both, whereas model (III) has neither. But model (II) disrupts this identification: on this model, there is an actual future history, although no history is privileged—all the histories are equally such that it is indeterminate whether that history is the actual future history (although some such history is the actual future history!). Needless to say, I am not claiming that this feature of model (II) makes perfect sense—it may not. (More on this below.) I am only claiming that this is what model (II) claims to be the case.

2.4 Semantics for *Will*

Thus far, we have simply put on the table three different models of the undetermined future. But now we must see how these three models interact with core topic of this book, viz., what *truth-value* should be assigned to future contingents—propositions of the form "It will be the case in n units of time that *p*", when some causally possible futures have it that *p* in n units of time, and some

[4] I certainly took no account of this view in my 2016a; in my more recent 2020, I simply identified "actual" and "privileged" (as have many others), and therefore described the model I am now calling model (II) as one on which there is a privileged branch, but it is indeterminate which branch is privileged. On reflection, however, this was perhaps not a sensible terminological choice: again, on model (II), there is a perfectly good sense in which *no* branch is privileged and they *all* have the *same* status (being such that it is *indeterminate* whether that branch is the actual branch).

do not. Here I wish to begin with what may seem to be a surprising fact, and that is that proponents of all three models can agree on what we might call a *neutral* semantics for "It will be in n units of time that *p*":

> (AAF) It will be in n units of time that *p* iff in all of the available futures, in n units of time, *p*.

More particularly, my proposal is that any statement of the form "It will be in n units of time that *p*" has the same truth-value (whether this is true, false, indeterminate, or something else) as the corresponding statement of the form "In all of the available futures, in n units of time, *p*". *Will* is thus treated as a universal quantifier over all *available* branches.[5]

Consider first model (I). On model (I), when we quantify over all of the available futures, how many futures is it that we quantify over? Well, the answer

[5] Thus, I defend a *modal* rather than a non-modal merely *temporal* semantics for *will.* (This distinction, and the literature surrounding it, is complicated; for the sake of simplicity and readability (and for the sake of not losing the forest for the trees), I am going to try to compress a lot of information into this note.) It is difficult to define the notion of a "modal". To call an expression a "modal" in the sense intended here is to say that its semantics involves, as Cariani and Santorio put it, the "manipulation of a world parameter" (2018: 132). As they explain (see also Kissine 2008), roughly, the literature on *will* is divided between merely temporal approaches, and modal approaches. According to a merely temporal semantics for *will, will* manipulates "exclusively a time parameter and no world parameter" (2018: 132). To see the contrast, we might state a merely temporal semantics as follows:

> "Will *p*" is true at a world *w* and a time *t* iff there is some time *t'* later than *t* and *p* is true at *w* and at *t'*.

A *modal* semantics for *will*, however, adds something to the temporal semantics. In particular, it shifts a world parameter as well as a time parameter. For example:

> "Will *p*" is true at a world *w* and a time *t* iff for all worlds *w'* that are open possibilities at *w* and *t*, there is some time *t'* later than *t* and *p* is true at *w'* and at *t'*.

And it is just this semantics that I defend for *will*, with the qualification that, for any world *w* and time *t*, the worlds that are open possibilities at *w* and *t* are those worlds that are consistent with the past of *w* up to *t* and the laws of nature of *w* and the future directed facts at *t* in *w*. In other words, the worlds that are open are the worlds that have futures that are "available" in the sense introduced above. Thus, strictly speaking, on the view I defend, *will* is a *combination* of a universal quantifier over worlds and an existential quantifier over times: roughly, *in all the available futures, there exists a later time such that p.* (I would also like to add "and is false otherwise" to the above statement of the modal semantics.) For an alternative modal semantics to the one just stated (on which *will* isn't a universal quantifier, but instead a "selectional modal"), see Cariani and Santorio 2018. One note: much of this book is concerned with how *will* interacts with negation. Certain issues about the scope of the negation with respect to *will* become simpler to address if we proceed in terms of a *metric* operator, rather than the "unrestricted" *will* operator; it is for this reason that, from the start, I ask about the truth-value of "It will be in n units of time that *p*", rather than simply "It will be [at some time or other] that *p*".

The motivations I propose for treating *will* as a modal are not precisely those to be found in the linguistics literature; still, it is common (perhaps even standard) in the linguistics literature to treat *will* as a modal (the primary question being which "worlds" are within the scope of the quantifier). For a start, see Enç 1996. For a dissenting voice, see Kissine 2008. (For instance, according to Kauffman (2005), *will* quantifies over the most "likely" futures, whereas, on Copley's (2009) proposal, *will* quantifies over the most "normal" futures.) The idea to treat *will* as a universal quantifier over some set of relevant futures is certainly not original to me; what is original to me, if anything is, is the proposal about how to understand which futures are relevant (this being those futures consistent with the past and the laws and the future directed facts (if there are any)).

is simple: one. In other words, for model (I), in order to determine whether in *all* of the available futures, in n units of time, we have *p*, we simply must check whether *the unique actual future*—viz., the sole available future—in n units of time has it that *p*. The point here is simple. For model (I), whether all of the available futures are *p* futures is always and everywhere a matter of whether the unique actual future is a *p* future. Compare. Suppose someone says that all of the people in the adjacent room are happy. Well, if there is only a single, unique individual in the adjacent room, what this person says is true just in case this single, unique individual is happy. So similarly here. If someone says that all of the available futures are *p* futures, and there is (always and everywhere) only one available future, what this person says is true just in case this single available future is a *p* future.

Intuitively, (AAF) gives us exactly what the Ockhamist wants. Consider the following:

(1) There will be a sea-battle in 24 hours.

According to my proposal, (1) means: In all of the available futures, in 24 hours, there is a sea-battle. Now we check. There is only one available future: the unique actual future. The question simply becomes whether the unique actual future is one in which there is a sea-battle in 24 hours. Indeed, observe at this stage that the proponent of model (I) will accept the following:

(UAF) It will be in n units of time that *p* iff in the unique actual future, in n units of time, *p*.

As I see it, however, that the Ockhamist accepts (UAF) is a consequence of something more fundamental: (UAF) is entailed by the truth of (AAF) together with the Ockhamist's contention that there are future directed facts which select for one unique future, and so only one future—the unique actual future—is available. The Ockhamist, however, can *agree* with a semantics on which *will* quantifies over all the available futures; again, the Ockhamist simply contends that (necessarily) only one future is available. A similar story holds for model (II). However, since, on this view, it is indeterminate which future is the unique actual future, it is (in the relevant cases) indeterminate whether *the unique actual future* is a *p* future. Accordingly, it is, in those cases, indeterminate whether it will be (in n units of time) that *p*.

Further, observe that (AAF), together with model (I) or (II), immediately predicts some results that are central to the issues to come—results having to do with how *will* interacts with *negation*. First, a very simple point. Consider the claim:

(2) All the cows in the barn are (such that they are) black.

Now, it is perfectly clear that the falsity of (2) does not amount to the truth of (3):

(3) All the cows in the barn are (such that they are) not black.[6]

Indeed, it is perfectly obvious that (2) and (3) could be false together—say, if there were two cows in the barn, one white, and the other black. But now note the following. Suppose we added (4):

(4) There is exactly one unique cow in the barn.

Now, (4) together with the falsity of (2) *does* entail (3). If it is false that all of the cows in the barn are black, and there is one and only one unique cow in the barn, then all the cows in the barn are not black. The point is simple. There is a sense in which the truth of (4) allows us to *bypass* the scope distinction between "Not all the cows in the barn are (such that they are) black" and "All of the cows in the barn are (such that they are) not black". (Much more on this to come.)

Now notice how these points relate to crucial issues concerning future contingents. Consider the following:

(5) All of the available futures are (such that they are) *p* futures (in n units of time).

= It will be in n units of time that *p*

(6) All of the available futures are (such that they are) not *p* futures (in n units of time).

= It will be in n units of time that ~*p*

[6] Note: I add the "such that they are" parenthetical to prevent a certain scope ambiguity; consider a standard (well-worn) example:

(g) All that glitters is not gold.

As everyone can see, the scope of the negation in (g) is ambiguous. The natural reading of (g) is of course:

(g*) Not all that glitters is such that it is gold.

But of course there is also the reading:

(g**) All that glitters is such that it is not gold.

Of course, a reading along the lines of (g**) must be forced—but it could be forced. (The boss at the gold mine, confused about the nature of gold, exclaims to his workers, "All that glitters is *not* gold—so give the glittery stuff a miss!") Examples could be multiplied, but in general, the point is simple: the scope of the negation in constructions like (g) is strictly speaking ambiguous. I bypass this issue in what follows by simply stating the points in terms of constructions like (g**).

(7) There is exactly one unique available future (and it extends for at least n more units of time[7]).

Note that the falsity of (5) together with (7) entails (6). If it is false that all of the available futures are *p* futures (in n units of time), and there is exactly one available future, then all of the available futures are not *p* futures (in n units of time). Now the point. (AAF) nicely gives us the result that, on models (I) and (II), there is a sense in which we can "bypass" the scope distinction between "Not all of the available futures are (such that they are) *p* futures"—what we might call the *negation of a prediction*—and "All of the available futures are (such that they are) not *p* futures"—what we might call the *prediction of a negation*. Further, if (7) is always and everywhere true—if it holds as a matter of metaphysics—then note that we will always therefore have the *disjunction* of (5) and (6), exactly inasmuch as we will always have it that the unique actual future (in n units of time) is a *p* future or is not a *p* future. And if (7) is assumed, note that it is not the case that both (5) and (6) can be false. Finally: if (7) is a pervasive non-semantic metaphysical assumption, then, pragmatically, we will tend to treat the falsity of (5) as equivalent to (6). However—and this is the crucial point—if (7) can be cogently denied on metaphysical grounds (that is, if model (III) can be defended), then we should end up with a view on which both (5) and (6) can be *false together*.

2.5 A Comparison: Cows in Barns

Perhaps the above points are clear enough. However, it is worth seeing if we can produce a sort of analogical comparison to make them yet clearer. As I see it, it is helpful to see that the logic of the situation here might be duplicated in other contexts in which debates about the "open future" are not at stake. And so here I return to the example above (suggested by Hughes 2015: 223) involving, well, cows in barns. Suppose there are two and only two cows in existence: one white, named Walter, and the other black, named Ben. We now consider the following claims:

(1) All the cows in the barn are (such that they are) white.
(2) All the cows in the barn are (such that they are) not white.

More particularly, we consider those claims given several competing hypotheses.

On hypothesis (I) (an analogue of model (I)), there is a unique single cow in the barn, and it is determinate which cow is the cow in the barn. The result is

[7] This parenthetical is an addition I shall subsequently supress for the sake of simplicity.

straightforward and simple. (1) is true and (2) is false or instead (2) is true and (1) is false. But certainly (1) and (2) aren't both false. And certainly the disjunction of (1) and (2) is true. One of the disjuncts is determinately true and the other determinately false—even if we may not know which.

On hypothesis (II) (an analogue of model (II)), there is a unique single cow in the barn, but it is indeterminate which cow is the cow in the barn. That is to say: either Walter and only Walter is in the barn (and so all the cows in the barn are white) or Ben and only Ben is in the barn (and so all the cows in the barn are not white (because black)), but—somehow!—it is indeterminate which cow is the cow in the barn. Thus, the disjunction of (1) and (2) is true—but... now we have to decide what to say.

There are perhaps various things we could attempt to say at this stage, but suppose we say this. Neither disjunct is true (both are indeterminate), and neither is false (again, both are indeterminate), but the disjunction *is* true. Of course, if this is what we say, we reach a puzzling state of affairs (one that famously rankled Quine): we have a true disjunction without either disjunct being true.[8] In defense of this stance, we reason as follows. There is a single cow in the barn. Well, suppose it is Walter. Then all the cows in the barn are white. But suppose it is Ben. Then all the cows in the barn are not white (because black). But then, since there is a single cow in the barn, *whichever cow it is*, the disjunction is true. The point: even if it is indeterminate which cow is in the barn, since, we have said, some single cow *is* in the barn, the disjunction of (1) and (2) is true no matter which cow is the cow in the barn. This remains the case even if it is not true (because indeterminate) that *Walter* is the cow in the barn, and not true (because indeterminate) that *Ben* is the cow in the barn.

On hypothesis (III) (an analogue of model (III)), both Walter and Ben are in the barn. Thus, there are multiple cows in the barn.[9] But then the situation once more is simple. (1) and (2) are both false. And here it is worth noting something crucial. What should we make of *truth no matter which cow is the sole unique cow in the barn*? We should think that that truth is no kind of truth at all.[10] Yes, if we *grant* that there is a sole unique cow in the barn, then certain things are going to come out true that we say are false. But we don't grant that there is a sole unique cow in the barn. Given that we don't grant that there is a sole unique cow in the barn, we shall therefore be unimpressed with the above reasoning in favor of the truth of the *disjunction* of (1) and (2). More particularly, there is something hypothesis (III) has in common with hypothesis (II): on both such views, it is not

[8] For a different approach (in terms of indeterminate truth), see Barnes and Cameron 2009.

[9] This is parallel to saying: there are multiple available futures. Or: in a two-branch scenario, that both branches are "available".

[10] Similarly: what should the proponent of model (III) make of the supervaluationist's notion of "supertruth", viz., *truth no matter which future is selected as the unique actual future*? They should say that that is no kind of truth at all.

true that all the cows in the barn are white, and not true that all the cows in the barn are black. On hypothesis (II), this is because both such claims are indeterminate; on hypothesis (III), it is because both such claims are false. But since hypothesis (II) maintains, whereas hypothesis (III) denies, that there is a sole unique cow in the barn, hypothesis (II) maintains that the relevant disjunction is true, whereas hypothesis (III) does not.

Now, the comparisons here are, I hope, straightforward, and I won't pause to note them all. I'll just note the following. Notice how proponents of hypotheses (II) and (III) both maintain that (1) and (2) are not true. In the same way, given (AAF), proponents of both models (II) and (III) can maintain the view—often taken to be definitive of the "open future"—that *future contingents are not true.* But there is, despite this commonality between models (II) and (III), nevertheless an important way in which proponents of model (II) may be inclined to join forces with proponents of model (I) to gang up on proponents of model (III). For note that model (III), together with (AAF), predicts the denial of what has been called Will Excluded Middle:

(WEM) It will be in n units of time that p or it will be in n units of time that $\sim p$.

And proponents of models (I) and (II) may be prone to say that WEM is not something we can reasonably deny. They may say, "There will be a sea-battle tomorrow or there will be no sea-battle tomorrow! How can you deny *that*?" (Again, much more on this to come.)

And here is where I need to tell a story. As it happens, a story involving cows in barns may help. Suppose that proponents of hypotheses (I) and (II) begin ganging up on the proponent of hypothesis (III). In particular, they begin shouting at that proponent in a common voice: "All the cows in the barn are white, or all the cows in the barn are not white!" What will the proponent of hypothesis (III) make of such shouting? The proponent of hypothesis (III) will remain serene in the face of such shouting—that is, in the face of the bald insistence that the disjunction of (1) and (2) must be true. She will remain serene for the following reason. She knows *why* this disjunction seems true to those who insist on its truth. It seems true to them because it *already* seems to them that there is one and only one cow in the barn. And there may be, in point of fact, an interesting further explanation of why they think there is one and only one cow in the barn. Perhaps proponents of hypotheses (I) and (II) are under the impression, having never seen the barn, that the barn is big enough for one and only one cow; accordingly, it has, since time immemorial, seemed to them that either all the cows in the barn are white or all the cows in the barn are not white, this being a consequence of the size of the barn and the nature of the cows. But our proponent of hypothesis (III) knows better—or thinks she knows better. Our proponent has, she thinks, finally seen the barn, and seen that it is plenty big for both cows—and, in point of fact, has seen both cows in the barn. Having seen both cows in the barn, she remains, again, serene in the face of the insistence that all the cows in the barn are white, or all the cows in the barn are not white. She knows why that seems like it must be true to the others. But she knows further that it needn't be true.

So similarly, I contend, for the relevant instances of "Will Excluded Middle". On the account I develop, we can know *why* it seems to so many that the relevant instance of "Will Excluded Middle" must be true. It seems like Will Excluded Middle is true (to those that it does) because it *already* seems to be true to them that there is a single available future—viz., the unique actual future. But there is, I contend, no such future at all. And once we see that, on metaphysical grounds, we can sensibly deny that there is a unique actual future, we can remain serene in the face of the bald insistence that Will Excluded Middle is a logical truth. Such is the story I aim to tell in this book.

2.6 Model (III) Once More

In this light, let us now return to the project of seeing how (AAF) interacts with the various models under discussion. And so let us now directly consider model (III). Again, (AAF) says

> (AAF) It will be in n units of time that *p* iff in all of the available futures, in n units of time, *p*.

As we have seen, on model (III), the available futures just are the causally possible futures. Thus, on model (III), we get the following result:

> (APF) It will be in n units of time that *p* iff in all of the causally possible futures, in n units of time, *p*.

Which is other words for saying: it will be in n units of time that *p* iff this result is causally *determined*. Now, here we must make the following crucial point. I defend (AAF), and I defend model (III). Together, these commit me to the truth of (APF). However, this does not commit me to the truth of (APF) as a *semantic* account of the *meaning* of *will*.

2.7 Peirceanism

If someone offers (APF) as a *semantics* for *will*—that is, as a proposal about its *meaning*—then what we have is familiar: it is the so-called (by Prior) "Peircean" semantics for *will*, according to which (roughly) *will* means "determined". Now, on its face, as a proposal about the *meaning* of *will*, Peirceanism seems implausible. The claim that there will be rain tomorrow does not seem to *mean* that it is causally determined for there to be rain tomorrow. Further, and relatedly, as various authors have pointed out, Peirceanism seemingly gets the wrong results when *will* claims are embedded under attitudes, e.g., *belief* and *fear*. Consider:

(j) John believes that there will be a sea-battle tomorrow.

The Peircean account predicts—wrongly—that John thereby believes that a sea-battle is determined for tomorrow. Further, consider

(jj) John fears that there will be a sea-battle tomorrow.

The Peircean account again predicts—wrongly—that John thereby fears that a sea-battle is determined for tomorrow. But surely John could fear that there will be a sea-battle tomorrow, without thereby fearing that this result is currently causally determined.[11]

[11] See, e.g., Hughes 2015: 226. For a related set of points, see Schoubye and Rabern 2017. See further Iacona's "Future Contingents" entry in the *Internet Encyclopedia of Philosophy*; Iacona notes that to hope that there will be peace tomorrow doesn't seem equivalent to hoping that it is determined that there will be peace tomorrow. Recent defenses of Peirceanism are few and far between; however, for one such defense, see Rhoda et al. 2006, and (relatedly) Rhoda 2007. For criticism, see Craig and Hunt 2013.

It is worth noting that Hartshorne was a "Peircean" long before Prior coined the term. Indeed, the following passage (from Hartshorne 1941) is, as best I can determine, the first time the view that future contingents are all false was ever put in print:

> It is sometimes argued, however, that we do know that future events are determinate. The law of excluded middle may be invoked. Either I will write the letter tomorrow or I will not write it tomorrow—only one of these can be true. The indeterminist may reply, Yes, only of them can be true, but perhaps both of them are false; for the truth may be that it is unsettled that I will write the letter, and equally unsettled that I will not. The proposition, "I will write the letter," is either true or false, but to say it is false is not to say that the proposition, "I will not write the letter," is true. For "I will do it" means that the present state of affairs (perhaps my resolution of will, in part) determinately excludes my *not* doing it, while "I will not do it" means that the present state of affairs excludes my doing it; but between these is the situation expressed by "I may or may not do it," which means that the present situation of myself and indeed the world in its totality is indeterminate with respect to my doing it. Or, in other words, it "will" occur means that *all* the possibilities for tomorrow which are still left open involve the occurrence in question; while it "may" occur means *some* of the open possibilities involve the occurrence in question; and it "will not" occur means that *none* of the possibilities involve it. Thus we meet once more the fundamental triad, the almost childishly simple but generally neglected mathematical key to philosophical problems, of all, some, and none. And no violation of the law of excluded middle as applied to propositions is in question. For surely to deny "all" is not to decide between "some" and "none." Hence if "it will occur" is the proposition *p*, then the corresponding negative or not-*p* is, not the proposition "it will not occur," but rather the following: "Either it will not occur or, at least, it *may not* occur." Hence, granting that, given any proposition *p*, either *p* or not-*p* is true (the law of excluded middle), it does not follow that the future is determinate. The only "middle" which indeterminism refuses to exclude is that between all (possibilities) and none, and this middle is universally admitted in logic. (1941: 100–101)

Notably, my account does considerably better with respect to these problems than does Peirceanism. My account predicts that, if John believes that there will be a sea-battle tomorrow, then John believes that all of the available futures have a sea-battle tomorrow. And that seems fair enough—although, of course, John himself may not express his belief in these (quasi-technical) terms. In other words, if John believes that there will be a sea-battle tomorrow, John believes that all of the ways things could be tomorrow that are consistent with the facts about how things will be tomorrow are ways in which there is a sea-battle tomorrow. So similarly for *fear*. If John fears that there will be a sea-battle tomorrow, John fears that all of the ways things could be tomorrow that are consistent with the facts about how things will be tomorrow are ways in which there is a sea-battle tomorrow.

The point here is this. I accept (APF) as *true*. However, I do not thereby accept (APF) as an account of the *semantic meaning* of *will*—that is, I do not thereby accept Peirceanism. Needless to say, of course, my account and the Peircean account nevertheless share a great deal in common—and face a great many of the same objections. My point is simply that they do not face all and only the same objections: the proposals are *semantically* importantly different. All in all: there is something right about Peirceanism. *Will* is a universal quantifier over future branches. And yet there is something wrong about Peirceanism: *will* is a universal quantifier over all *available* branches, and whether the *available* branches are one and the same as the causally possible branches is a separate (non-semantic) matter entirely.

Thus, I reject any argument to the effect that, because I endorse (APF), my proposal somehow "reduces to" Peirceanism.[12] The core idea that I need to block such a reduction is the widely endorsed idea that two propositions can be logically equivalent (necessarily have the same truth-value as one another), and yet differ in meaning. And that is my proposal. My proposal is that "It will be in n units of time that *p*" necessarily has the same truth-value as "In all of the causally possible futures, in n units of time, *p*", but necessarily has the same truth-value as *and means* "In all of the available futures, in n units of time, *p*". This is because, as I see it, necessarily, there are no primitive future directed facts, and so, necessarily, the available futures are the causally possible futures. However, this latter contention is a sheer metaphysical contention about which I could be entirely mistaken, independently of the semantics I have suggested for *will*. Or look at it this way. I accept (AAF) and (APF) as both *true*. However, I accept (AAF) on *semantic* grounds (on my view, it is an *analytic* truth), whereas I accept (APF) on metaphysical grounds (it is *not* an analytic truth).

[12] Cf. both Schoubye and Rabern 2017 and Wawer 2018, who object to my earlier 2016a ("Russellian") proposal on just these grounds.

Consider, then, the following statement from Hughes: "So for [the] Peircean . . . , *p* means that—or at any rate 'comes to' (necessarily has the same truth-value as, is logically equivalent to) *it is inevitable that p*" (Hughes 2012: 46). But my contention is that we must be careful in this context precisely to distinguish between the different notions identified here by Hughes. Again, it is open to an open futurist to maintain that *it will be that p* is *logically equivalent*—necessarily has the same truth-value as—*it is inevitable that it will be that p*, without maintaining that these claims have the same *meaning*. Note: on standard assumptions about God (assumptions discussed in Hughes' own paper), *p* will be logically equivalent to *God believes p*. (See Chapter 6.) But it certainly doesn't follow (to pick up on one of Hughes' worries for the Peircean) that someone who fears that he has been fired from his job thereby fears that *God believes* that he has been fired from his job. The point: if we maintain, as we may very well maintain, that *p* is logically equivalent to *God believes p*, then we will want some way of maintaining that two propositions can be logically equivalent yet differ in meaning. And once we do that, we can maintain that *it will be that p* and *it is inevitable that it will be that p* are logically equivalent, but differ in meaning. And once we see *that*, we can see that many standard objections to "Peirceanism"—as a claim about the *meaning* of *will*—are not thereby cogent objections to the logical equivalence noted above.

With these points on the table, we can also observe that one key objection to Peirceanism does not apply to the view I defend in this book. Notably, Prior observed that, on Peirceanism, future contingents are "perversely inexpressible" (1967: 130). Similarly, Correia and Rosenkranz write: "However, Peircean accounts are notoriously impoverished, as they leave us with no means at all to express the thought that something will be the case as a matter of mere historical contingency. In other words, we could not even formulate future contingents" (2018: 102). As I see it, this problem for the Peircean arises precisely because the Peircean does not make a semantic distinction between the *available* futures and the causally possible futures. But first, let us see if we can unpack slightly more carefully the objection articulated here by Correia and Rosenkranz, as I understand it. At least as a first pass, we might try defining a "future contingent" as follows:

> A future contingent =df. a claim that it will be that *p*, when it is not the case that all causally possible futures feature *p*.

But for the Peircean, this amounts to saying:

> A future contingent =df. a claim that all causally possible futures feature *p*, when it is not the case that all causally possible futures feature *p*.

And thus future contingents contain an *internal contradiction*—and thus we cannot even *express* the thought that a future contingent might be true.

However, contra the Peircean, the idea that something could be such that it will happen "as a mere matter of historical contingency" is not, in itself, a contradictory idea. It is not *analytically* false that something could be such that it will happen as a matter of mere historical contingency—but Peirceanism says that it is.

It is worth seeing how my view does not suffer from a similar problem. We can, on my view, keep the above definition of a future contingent. On the view I favor, however, this amounts to saying:

> A future contingent =df. a claim that all available futures [all causally possible futures consistent with the primitive future directed facts, if there are any] feature *p*, when it is not the case that all causally possible futures feature *p*.

And this much is certainly no semantic contradiction; thus, it is certainly possible, on my view, to *express* future contingents. In other words, on my view, it is not *analytically* false that there are true future contingents. Again, on the view I favor, there are certainly no true future contingents, but this isn't because the mere idea of a true future contingent is an internal contradiction—it is because, *metaphysically*, there are no primitive future directed facts.

2.8 All False

The core argument of this book can now be stated as follows.

> Semantically, *will* is a universal quantifier over all available branches.
> Metaphysically, there are no primitive future directed facts, and so the available branches just are the causally possible branches. (model (III))
>
> Result: *future contingents are all false.*

For if "It will be in n units of time that *p*" quantifies over all the available branches, and those *just are* the causally possible branches, and the given claim is a future *contingent*, then since it is just plain false that, in the case of a future contingent, *all* the causally possible futures feature that *p* (in n units of time), then future contingents will simply be false.

2.9 A Flow-Chart

Perhaps it will be helpful to have something of a pictorial representation of the options discussed in this chapter, as I am conceiving of them. Consider Figure 2.1.

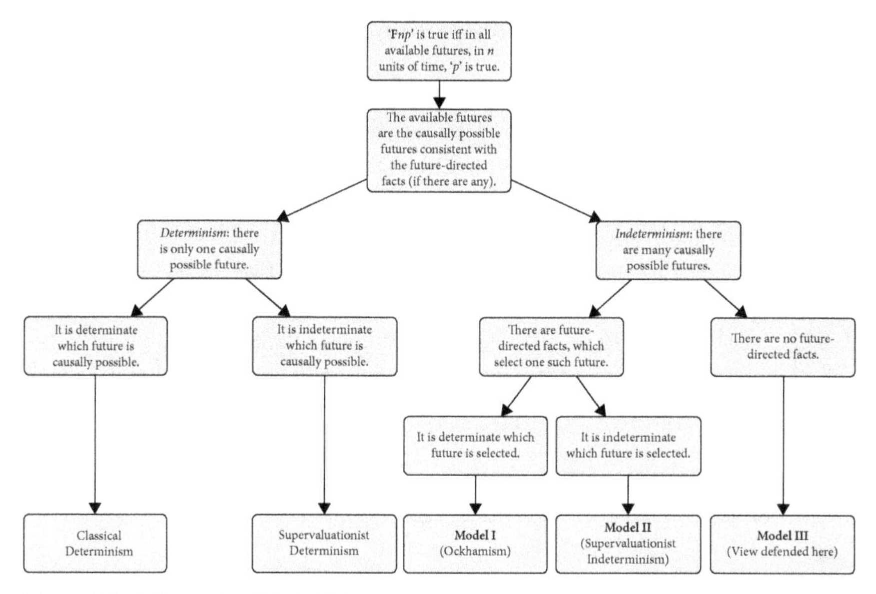

Figure 2.1 A flow-chart for 'will'

Let me make several brief comments about this chart. First, note that this chart represents how I think we should think about the option space, once the first two points on the chart are granted. (Those who deny either of these assumptions will not find their view represented in Figure 2.1.) Second, note that I have only discussed those options that assume indeterminism.[13] Third, observe that, on the first four options (along the bottom row), there is a sole *available* future—the unique actual future—and observe that those options validate WEM. Hence the core observation: the status of WEM is tied to whether there exists a unique actual future. If there is no unique actual future, WEM fails. Finally, observe that my own position is the consequence of the initial two broadly "semantic" assumptions, together with two *metaphysical* assumptions: there are many causally possible futures, and no primitive future directed facts to discriminate between them.

[13] For an articulation of what I have here labeled "supervaluationist determinism", however, see Barnes and Cameron 2009. Barnes and Cameron suggest (without endorsing) a fascinating view on which though there is only one future consistent with the past and the laws—and so determinism is true!—it is metaphysically indeterminate which future that is. The result, they maintain, is that there is a sense of "openness" on which determinism is consistent with the "openness" of the future. If it is enough for the future to be "open" that it should be indeterminate what is going to happen, then this is a view on which the future is open, precisely because, on this view, it is indeterminate what is causally determined. (Here it is worth re-emphasizing that I do not purport to "own" the terminology of "openness"—I have stipulated that, for me, only models II and III are models in which the future is "open", but, again, this is merely terminological.)

2.10 Objection: Missing Ambiguities?

I now wish to address one objection that *may* have occurred to some readers—and one that allows us to set up some of the issues to come. (Anyway, the objection needs to be addressed, and I haven't found a better place to address it than here.) Note that my overall argument contains two essential elements: a *semantic* claim, and a *metaphysical* claim. The objection I want to consider pertains to my *semantic* proposal to treat *will* as a universal quantifier over available branches. At this stage, one might object that, if *will* were such a universal quantifier, then we should expect to see *scope ambiguities* concerning how it interacts with negation—indeed, precisely the sort of scope ambiguities we see in sentences like (g) (which can be read as either (g*) or (g**)):

(g) All that glitters is not gold.
(g*) Not all that glitters is such that it is gold.
(g**) All that glitters is such that it is not gold.

The thought, then, is that if *will* is a universal quantifier (over available branches), then there should be *two* places for the negation in (9) to appear with respect to the *will*:

(9) There will not be a sea-battle in one hour.

(9a) Not all the available futures are futures in which there is a sea-battle in one hour.

(9b) All the available futures are futures in which there is not a sea-battle in one hour.

The objection becomes acute if we maintain—as per model (III)—that the available futures just are the causally possible futures. After all, certainly (9) doesn't seem like it could be true merely because (as in (9a)) *not all* the causally possible futures are futures in which there is a sea-battle in an hour. For that is merely to say that a sea-battle isn't determined to take place in an hour. A sea-battle isn't determined to take place in an hour. So there will not be a sea-battle in an hour. But how could *that* follow?[14]

[14] For one statement of this objection (as it applies to Peirceanism), see De Florio and Frigerio 2019: Ch. 3. See further Schoubye and Rabern 2017. The objection is similar in spirit to one of Stalnaker's objections to Lewis' semantics for the counterfactual; as discussed later, Lewis treated the counterfactual conditional as a sort of universal quantifier over a certain class of worlds. Hence Stalnaker (1981: 93): "[On Lewis' view] to assert 'if A, then B' is to assert that B is true in every one of a set of possible worlds defined relative to A. Therefore, if this kind of analysis is correct, we should expect to find, when conditionals are combined with quantifiers, all the same scope distinctions as we find in quantified modal logic." Stalnaker, of course, maintained that those distinctions can't be found. I investigate

What we need here is thus either

(a) a story about how, in constructions like "there will not be a sea-battle in one hour," the negation can never outscope the universal quantifier expressed by *will* (i.e., such sentences always mean (9b), and never (9a))

(b) evidence that a sentence like "there will not be a sea-battle in an hour" can be true so long as *some* available future is a non-sea-battle future in one hour (i.e., that such a sentence is indeed ambiguous between (9a) and (9b)).

Either (a) or (b) would do. For the sake of completeness, however, I investigate both.

First, observe that there are certain "modals" that are very plausibly universal quantifiers that do not also give rise to the relevant scope ambiguities, when that modal is followed immediately by negation.[15] Consider:

(8) There must not be a war in Australia right now.

Now, (epistemic) *must* is plausibly a universal quantifier of some kind—namely, a universal quantifier over all of the epistemically available scenarios (which says that the given content holds in *all* such scenarios). So should it then follow that (8) should be strictly ambiguous between the following?

(8a) Not all epistemic possibilities are possibilities in which there is a war in Australia.

(8b) All epistemic possibilities are possibilities in which there is not a war in Australia.

Well, why should it? Intuitively, (8) means (8b), and cannot mean (8a). (8) means that it is certain that there is no war in Australia, not that it is not certain that there *is* a war in Australia.[16] And so similarly for *will*. Just because *will* is analyzed as a universal quantifier, it does not follow, by itself, that we should therefore expect a scope ambiguity in (9) (between (9a) and (9b)). Intuitively, (9) means (9b), and cannot mean (9a). (9) means that all the available futures are ones in which there

quantifiers in the chapter to come—but plausibly Stalnaker could just as well have said "when conditionals are combined with negation . . . ".

[15] Thanks to Wolfgang Schwarz for suggesting this point.

[16] A similar observation holds for *should*—which is (see Ch. 3) plausibly also a "neg-raiser". Consider a claim like

(8*) John should not go to Australia.

Intuitively, (8*) does not have the reading that it is merely *not the case* that John should go to Australia; (8*) means that John should refrain from going, and cannot merely mean that it is not the case that he should go.

fails to be a sea-battle in an hour, not that it is not the case that all such futures are futures with a sea-battle in an hour.

Or does it? In point of fact, the case that (9) cannot mean (9a) is open to serious doubt. And so now let us investigate option (b). Notably, there is important precedent for the claim that (9) *can* mean (9a). In a neglected passage in Chapter 7 of his 1967 *Past, Present, and Future*, Prior writes:

> Turning now to the other way of answering the argument [for fatalism] . . . , I begin by modifying the ancient and medieval presentation of this alternative at one point. What is said by writers like Peter de Rivo is that predictions about an as yet undetermined future are neither true nor false. It did seem to me in the early 1950s that this was the only way to present an indeterminist tense-logic, but in *Time in Modality* two alternatives to this were mentioned, one . . . which I now want to pursue further. What here takes the place of a third truth-value is a sharp distinction between two senses of 'It will not be the case the interval *n* hence that *p*'. This may mean either
>
> (A) 'It will be the case the interval *n* hence that (it is not the case that *p*)', i.e. *FnNp*;
> or
> (B) 'It is not the case that (it will be the case the interval *n* hence that *p*)', i.e. *NFnp*.
>
> 'Will' here means 'will definitely'; 'It will be that *p*' is not true until it is in some sense *settled* that it will be the case, and 'It will be that not *p*' is not true until it is in some sense settled that not-*p* will be the case. If the matter is not thus settled, both these assertions, i.e. *Fnp* and *FnNp*, are simply false . . . There is now no question of denying the Law of Excluded Middle . . . and moreover the allied metalogical 'Law of Bivalence' . . . is not abandoned either.[17]

Unfortunately, however, Prior does not provide any evidence that such "wide scope" readings (as in (B)) are possible. Can any such evidence be provided? Arguably, evidence that wide scope readings of the negations in "will not . . . " constructions are *possible* is given by the fact that such constructions admit so-called "negative polarity items"—especially "yet". Consider the following news headline penned by Chris Hall from Canada's CBC news on December 4, 2017: "Free trade talks with China will not yet begin".[18] Now, one might initially wonder

[17] Prior 1967: 128–129. Note: I defend Prior's result, without any stipulation that 'will' means 'will definitely'—that is, without assuming Peirceanism.

[18] http://www.cbc.ca/news/politics/trudeau-china-meeting-1.4431136 (accessed February 20, 2018).

whether what is being said here is that though free trade talks with China *will* begin, they just will not begin *yet*. But this is not so. The story reads as follows:

> Canada will not begin formal free trade talks with China—at least not yet.
>
> Prime Minister Trudeau met Monday with Chinese Premier Li Keqiang at the Great Hall of the People in Beijing.
>
> A joint statement indicating talks would begin had been widely expected.
>
> But the first clues that it might not happen emerged late in the day.

Now the plain reading of the first sentence of this story is one on which the negation is taking wide scope with respect to the *will*. More particularly, the continuation "at least not yet" is both felicitous *and* forces a wide scope reading of the negation—a reading that says, in short, that it is not *yet* the case that it will be the case that trade talks with China begin (and here it is meant begin *at all*). Other examples could be provided; herewith one from Tolkien:

(h) No, my heart will not yet despair. Gandalf fell and has returned and is with us.

But if *that* seems fine, then so does

(h*) No, my heart will not despair—at least not yet.

Now, *prima facie*, someone saying "My heart will not despair" should be understood as saying that, for all available histories, at no point during those histories does her heart despair. (Recall that if there is a unique actual future history, this is tantamount to saying that, at no point in the actual future history does one's heart despair.) But the continuation "at least not yet" forces a different interpretation of this utterance. After all, plainly it makes no sense to say:

> For all available future histories of the world, at no point in those histories does my heart despair—at least not yet.

Similarly, if we suppose that there is just one available future history (the unique actual future history), note that, again, it makes no sense to say:

> At no point in the actual future history of the world does my heart despair—at least not yet.

After all, *ex hypothesi*, what obtains in the actual history of the world cannot be *added to* in this manner; what happens in the actual history of the world is fully specified. The point: "My heart will not despair" is strictly speaking ambiguous.

Are you saying that it will always be that your heart does not despair—or are you merely saying that your heart will not *yet* despair, that is, that it is not *yet* the case (but may *become* the case) that your heart will despair? In other words, are you *predicting* that it will always be that one's heart does not despair—or are you instead simply *denying* that it is now already such that it will despair? Do we have here the prediction of a negation (over the relevant length of history)—or instead merely the negation of a prediction?

Return now to our central example: "There will not be a sea-battle in one hour". And consider a context in which Jones is convinced that a sea-battle in one hour is a foregone conclusion. He says, "There's no use: there will be a sea-battle in an hour." Believing the matter still to be open, Smith says:

(17) There will not be a sea-battle in an hour—at least not *yet.*

(17)—like the first line of the story above—is uncommon but (perhaps) acceptable. But someone asserting (17) means to assert

(17*) There will not *yet* be a sea-battle in an hour.

Which is to say:

(17**) It is not yet the case that there will be a sea-battle in an hour.

Which is to say that it is possible that the negation in (9) could be forced to take wide scope—exactly as Prior maintained.

Well, where are we? Recall the objection: treating *will* as a universal quantifier predicts that (9) should be ambiguous between (9a) and (9b), but that it is absurd to say that (9) could be true so long as (9a) is true. I have considered two avenues of response to this objection. The first is simply to deny that this ambiguity is predicted at all—that is, we simply deny that treating *will* as a universal quantifier requires seeing some ambiguity in (9) in the first place. Such a result does not hold for other modals such as *must*—so why must it hold for *will*? The second is to simply defend exactly this ambiguity: (9) *is* ambiguous (as Prior maintained) between (9a) and (9b). In the end, I am officially agnostic on which approach is the right one. However, I do wish to note the following. Certainly it seems that the *default* reading of "will not..." constructions is one on which the negation is taking narrow scope with respect to *will.* In what follows in this book, I will, in general, ignore the possibility of readings on which it is instead taking wide scope. (In the next chapter, however, we will want to remember how, if (9) is ambiguous, this complicates in some ways my approach to how future contingents interact with the classical Law of Excluded Middle.)

2.11 Costs and Benefits: A Preliminary Take

It is worth quickly taking stock of the costs and benefits of each of our respective models.

Recall model (I). On model (I), in an indeterminist setting, there is a privileged future history. Benefits: proponents of model (I) needn't make sense of the "indeterminacy" at issue in model (II), and needn't address the semantic and logical issues that face model (III). The costs, however, are simple. On model (I), we cannot accept the core intuition that in an indeterminist setting, no history is privileged over the others. The cost is ontological: we must posit a privileged branch.

Recall model (II). On model (II), in an indeterminist setting, there is no privileged future history, albeit there is an *actual* future history—it is just indeterminate which future history is the actual one. Benefits: we respect the idea that in an indeterminist setting, no branch is *privileged*, but because we accept that there is an *actual* future history, we can accept Will Excluded Middle, thus avoiding at least some of the difficulties inherent in model (III). The costs, however, are that we *cannot avoid ontological commitment to a unique actual future.* (More on this to come.) Further, unlike as in models (I) and (III), we must make sense of the relevant "indeterminacy". And some, at least, have expressed doubts about whether this can be done. Consider, for instance, the following objection from Hughes:

> To start with, according to the Thomasonian egalitarian [the open futurist supervaluationist]
>
> > There are many still possible histories, and there is just one actual (still possible) history, but there is no fact of the matter about which still possible history is the actual (still possible) history. (That is, for no still possible history is it either true or false that that still possible history is the actual (still possible) history.)
>
> But does this make sense? It sounds worrisomely like:
>
> > There are many cows, and there is just one cow in that barn, but there is no fact of the matter about which cow is the cow in that barn. (That is, for no cow is it either true or false that that cow is the cow in the barn.)
>
> ...[S]uppose we think (plausibly enough) that it is impossible that there are many cows, and there is just one cow in that barn, but there is no fact of the matter about which cow is the cow in that barn, and likewise impossible that there are many still possible histories, and there is just one actual (still possible) history, but there is no fact of the matter about which still possible history is the actual (still possible) history. Then we'll think: if a certain (Thomasonian) account of truth-at-a-time-on-a-history and truth-at-a-time allows it to be true

> that there is just one actual (still possible) history, but there is no fact of the matter about which still possible history is the actual (still possible) history, so much the worse for that account. (Hughes 2015: 223–224)

I am inclined to agree with Hughes concerning this assessment, but I shall not press the point. My goal in this book is not to argue that no sense can be made of this kind of indeterminacy; my goal is to argue that, at least when it comes to the open future, we do not *need* to invoke this kind of indeterminacy.

Recall model (III). On model (III), in an indeterministic setting, there is no such thing as "the actual way things will go"—that is, no such thing as the actual future history. Benefits: model (III) needn't make sense of the "indeterminacy" of model (II), and clearly avoids ontological commitment to a privileged branch (as in model (I)) and an actual branch (as in model (II)). The costs associated with model (III) are varied, and explored in the chapters to come. My claim, again, is simply that we have no *need* to appeal to an actual future. However, by way of introducing the problems that dominate the next several chapters, let me try to motivate why some have thought they have seen such a need. For my view is that there is *no* actual future history. And this, it seems, can appear to lead to some uncomfortable results. Consider, from the perspective of a proponent of model (II), an objection as follows:

> We are speaking in an indeterministic context, one in which a sea-battle tomorrow is objectively possible, and in which there being no sea-battle tomorrow is objectively possible. But these are the only two possibilities for tomorrow. Accordingly, we are in a going-to-be-a-sea-battle-tomorrow context, or instead a going-to-be-no-sea-battle context. It is just *indeterminate which* context we're in. But it isn't as if we are in *no* such context. We indeed are in one such context, but it is, again, just indeterminate which context we're in. But this is to say that there *is* an actual future history—a complete context we're in—but it is just indeterminate which future is actual; that is, which branch indeed represents the complete context we're in. And we *can* speak of "the results of the indeterministic process" at issue—it is just currently *indeterminate* to what this description refers. It refers to something, but it is indeterminate to what.
>
> Your view, however, would have it that we are not in a sea-battle-tomorrow context, and not in a peace-tomorrow context. That, however, is the wrong thing to say. Consider the sea-battle-tomorrow context. Is that the context we're in? Your answer: no. There is, you say, no actual branch. If so, then the sea-battle branch is not the actual branch, which is to say that it is not the context we are in. So we are not in a sea-battle-tomorrow context. However, if the future is *open* with respect to sea-battles tomorrow, it is much too strong to say that we are not in a context in which there is going to be a sea-battle tomorrow. For to say that we are not in a context in which there is going to be a sea-battle tomorrow is

> tantamount to saying that there is not going to be a sea-battle tomorrow—which is in turn tantamount to saying that there is going to be no sea-battle tomorrow. But if the future is open with respect to sea-battles tomorrow, that is plainly something that we just can't say.

This is a seductive argument, and the bulk of Chapters 3–6 can be read as an attempt to answer it. My answer is as follows. That we are not in a context in which there is going to be a sea-battle tomorrow does *not* mean that we are *in* a context in which there is going to be *no* sea-battle tomorrow. We are not in a context in which there is going to be a sea-battle. But we are also not in a context in which there is going to be no sea-battle. We are in neither such context. (But we *are* in a context in which it is going to be that there is a sea-battle or no sea-battle tomorrow.) From the fact that we are not in a context in which there is going to be a sea-battle, it would follow that we *are* in a context in which there is going to be *no* sea-battle only if there *already* is something like the complete context that we're in, future facts included. In that case, of course, since our context is complete, if it doesn't feature a sea-battle tomorrow, it features the absence of a sea-battle tomorrow. That is just what it is for it to be *complete*. But whether we are in such a context—whether there is indeed a unique actual future history, an "actual world"—is precisely what is at issue.

Well, this is what is at issue, perhaps, in various *philosophical* discussions. Ordinary thought and talk, however, is not necessarily sensitive to metaphysical arguments to the effect that there is no actual future history, no such thing as "the complete context we're in". It is for these reasons, I contend, that *saying* "There is not going to be a sea-battle tomorrow!" *sounds* too strong, on the assumption that it is *open* that there should be a sea-battle tomorrow. But what can sound too strong can often nevertheless be strictly speaking true.

Here is an example. Jack has just been invited to Jill's theme party—a party he regards, in his words, as a stupid party. But nevertheless, he's on the fence. For, he reasons, he sort of likes stupid parties. Jill asks Jack about his plans. Jack says, "Jill, I don't want to go to your stupid theme party." Jack, however, is on the fence about going—or so we had thought. Yet Jack's statement sounds much too strong to be consistent with Jack's neutrality. But Jack insists that it is, and that it is true. For, he says, he doesn't want to go to Jill's stupid party—but he doesn't want to *not* go to Jill's stupid party either. Both claims—that he wants to go, and that he wants to not go—are *false*. As we said: he's on the fence.

My project is now to argue that the open future is much the same.

3
The Open Future, Classical Style

Various philosophers have long since been attracted to the doctrine that future contingent propositions systematically fail to be true—what is sometimes called the doctrine of the *open future*. However, open futurists, in this sense of the term, have always struggled to articulate how their view interacts with standard principles of classical logic—for instance, with the Law of Excluded Middle (LEM). For consider the following two claims:

There will be a sea-battle tomorrow
There will not be a sea-battle tomorrow

According to the kind of open futurist at issue, both of these claims may well fail to be true. According to many, however, the disjunction of these claims can be represented as $p \vee \sim p$—that is, as an instance of LEM. And if this is so, the open futurist is plainly in a difficult position. She must either simply deny LEM outright, or instead maintain that a disjunction can be true without either of its disjuncts being true. And whereas open futurists have defended both such options with considerable care and ingenuity, both are also faced with substantial costs.[1]

In this chapter, however, I wish to explore a different option—an option that has been articulated and defended by at least one of the leading lights of tense-logic in the twentieth century (A.N. Prior) but is nevertheless often bypassed and even ignored. This is the position that, in fact, the disjunction of the above two claims *cannot* be represented as an instance of $p \vee \sim p$. And this is for the following reason: the latter claim is not, in fact, the strict negation of the former. More particularly, there is an important semantic distinction between the strict negation of the first claim [~(There will be a sea-battle tomorrow)] and the latter claim (There will not be a sea-battle tomorrow). And LEM, of course, is concerned with strict negations.[2] If this semantic distinction can be maintained, the open futurist's prospects concerning LEM appear much more hopeful. For starters: she can maintain that neither of the above claims is true, but that this, in itself, tells us nothing about LEM—because the disjunction of these claims is not an instance of

[1] For the former approach, see Lukasiewicz 1920; for the latter, see esp. Thomason 1970.
[2] For differing defenses of this position, see Prior 1957: 95–96 and 1967: 128–129, Hartshorne 1941: 100–101 and 1965, Bourne 2006: 91, Rhoda et al. 2006, Seymour 2014, Wilson 2016: 114, Hess 2017, and Todd 2016a and 2020.

The Open Future: Why Future Contingents are All False. Patrick Todd, Oxford University Press. © Patrick Todd 2021.
DOI: 10.1093/oso/9780192897916.003.0004

LEM. More to the point: she can maintain that when we *do* have the strict negation of the first claim, we have a claim that is just plain *true*, even if the future is "open". If the future is open regarding sea-battles tomorrow, then currently *it is not the case that* there will be such a sea-battle tomorrow. But this isn't to say that there *won't* be a sea-battle tomorrow. Accordingly, the open futurist can maintain that the *real* instance of LEM is just plain true. Problem solved.

Or so it seems. Again, the possibility of this approach is more often ignored than it is denied. But it is sometimes denied. In particular, it has been denied by Cahn (1967: 60–65), Thomason (1970), and more recently by John MacFarlane (2014) and Fabrizio Cariani and Paolo Santorio (2018), the latter of whom call the denial of the given semantic distinction "scopelessness". According to these authors, that is, *will* is "scopeless" with respect to negation; for instance, whereas there is perhaps a *syntactic* distinction between 'It is not the case that it will be tomorrow that there is a sea-battle' and 'It will be the case tomorrow that it is not the case that there is a sea-battle', there is no corresponding *semantic* distinction—that is, no corresponding difference in *meaning*. And if this is so, the approach in question fails.

In this chapter, however, I defend the unorthodox position that the above two claims are classical *contraries* (both can be false, but not both true)—and I thereby criticize the claim that *will* is "scopeless" with respect to negation. The central theme underlying my position is this: philosophers (and semanticists) are mistaking semantic competence with *will* with what is in fact semantic competence with *will* together with *an implicit metaphysical model of the future*—one on which there is a unique actual future. (See models (I) and (II) in the previous chapter.) However, once this model is denied, the judgments undergirding scopelessness lose their motivation and justification. Thus, in this chapter, I develop a sort of "error-theory" for certain ordinary semantic intuitions about *will*. Adopting Prior's (1957: 11–12) metric tense operator 'Fn*p*' as shorthand for 'It will be in n units of time hence that *p*', I contend that the dominant metaphysical model of the future implicit in ordinary, unreflective discourse renders it the case that ~Fn*p* implies Fn~*p*. Accordingly, when this model is operative in the background of the discourse, it is, naturally, unimportant to distinguish between ~Fn*p* and Fn~*p*, for according to this model, whenever we have the former, we have the latter. However, once one firmly denies the given model of the future, all bets are off: we can see that ~Fn*p* does not *mean* Fn~*p*. The distinction, to be sure, is a philosophers' distinction. But it is a distinction nonetheless. And when it comes to articulating the *philosophical* theory of the open future, this is what matters.

I develop this point by defending the claim that *will* is a so-called *neg-raising predicate*. "Neg-raising" refers to the widespread linguistic phenomenon whereby what is in fact semantically wide-scope negation gets treated, in context, as if it belonged to the relevant embedded clause. For instance, *I don't think that Trump is a good president* strongly tends to implicate *I think that Trump is not a good president*—despite the former not semantically entailing the latter. The phenomenon of neg-raising has generated a substantial discussion in the linguistics literature—and the present chapter thus aims, in part, to make a contribution to that literature. However, I do not aim to make a direct contribution to providing a *theory* of neg-raising—a theory that would predict which predicates are neg-raising and why. Rather, my aim here is to show that, whatever its fundamental explanation, the phenomenon can also be seen to apply to *will* and *will not.* As I hope becomes clear, seeing *will* as a neg-raiser promises to solve what would otherwise be intractable problems in the philosophical theory of the open future.[3]

3.1 Scopelessness

As noted above, some authors have claimed that *will* is "scopeless" with respect to negation. Under the heading *Missing Scope Distinctions*, John MacFarlane articulates the thesis by comparing two claims, and writes as follows:

> (13) It is not the case that it will be sunny tomorrow
> ~ Tomorrow *S*
> (14) It will be the case tomorrow that it is not sunny
> Tomorrow ~*S*
> It is striking, though, that although we can mark the syntactic distinction by resorting to cumbersome circumlocutions, as in (13)–(14), these variants seem like different ways of saying the same thing. (2014: 216)

According to MacFarlane, then, we can simply push the negation in (13) inside the scope of the "Tomorrow" (thereby getting (14)), and we can do so without any change in meaning.[4] Now, MacFarlane makes the claim that the requisite scope distinctions are "missing" in the context of a criticism of the so-called (by Prior) "Peircean" semantics for *will*, which treat *will* as a universal quantifier over all open causal possibilities. (These are the semantics defended in Prior 1957: 95–96, 1967: 128–129, and Hartshorne 1965.) If 'It will be in n units of time that *p*' meant 'On all causally possible branches, in n units of time hence, *p*', we would expect to see a plain scope difference between ~Fn*p* and Fn~*p*—which we don't see,

[3] Indeed, my aim here is to provide just the machinery alluded to in Prior's apt remark, "I cannot help suspecting that the theory of neuter propositions arose out of insufficient machinery to distinguish between ~Fn*p* and Fn~*p*." Prior 1967: 136.

[4] Cf. also Craig's (1987: 62) comments on Prior's "Peirceanism": "But does such a reinterpretation make any difference at all? To say that it is not the case that Bush's election will be the case seems to be the same as saying that Bush's election will not be the case."

according to MacFarlane.[5] Notably, however, essentially the same objection would seem to apply to the account I proposed in the previous chapter—an account on which *will* is a universal quantifier over all *available* branches.

Following MacFarlane, Cariani and Santorio write:

> Our second constraint [on developing a semantics for *will*] is that *will* is scopeless with respect to an important class of other linguistic items. By this we mean that changes in the relative syntactic scope between *will* and these other items don't make a difference to the truth conditions of *will*-sentences. This is a remarkable feature of *will*, and one that is not generally shared by modal expressions. For present purposes, it is enough to observe scopelessness with respect to negative items, as illustrated by:
>
> (9) a. It will not rain.
>
> b. It is not the case that it will rain.
>
> (9)a and (9)b are truth conditionally equivalent...In short, *will* appears to commute freely with ordinary English negation.... The lack of scope interactions with negation immediately yields an interesting logical constraint:
>
> Will Excluded Middle (preliminary take): 'Will A $\vee$ Will ~A' is a logical truth.[6]
>
> (2018: 134–135)

Neither MacFarlane nor Cariani and Santorio put their points in terms of a *metric* tense operator; MacFarlane employs a (similar) 'tomorrow' operator, and Cariani and Santorio employ a non-metric 'will' operator. For various reasons, however, it will be convenient in what follows to employ a metric operator—and clearly whatever reasons MacFarlane and Cariani and Santorio have given above for "scopelessness" regarding 'tomorrow' and 'will' apply *mutatis mutandis* to 'It will be in units of time hence that...'. Thus, again, adopting 'Fn*p*' as shorthand for 'It will be in n units of time hence that *p*', the claim at issue is that there is no semantic distinction between ~Fn*p* and Fn~*p*. According to these authors, that is, making a sharp distinction between ~Fn*p* and Fn~*p* is approximately similar to making a sharp distinction between $(p \wedge q)$ and $(q \wedge p)$. The "order" of the negation with respect to *will* is as semantically irrelevant as is the "order" of the

[5] A similar argument is offered by Hughes 2012: 48.

[6] As Cariani and Santorio recognize, if a given operator **S** is scopeless, then if the Law of Excluded Middle is a logical truth, then **S**-Excluded Middle will be a logical truth, viz. '**S***p* $\vee$ **S**~*p*'. Assuming LEM, the claim that an operator **S** is scopeless is, therefore, equivalent to the claim that **S**-Excluded Middle is a logical truth. However, it is worth noting, as I note again shortly vis-à-vis- MacFarlane –that the claim that Will Excluded Middle is a *logical* truth is immediately complicated by the observation that a logical truth should be true at the last moment of time—but Will Excluded Middle is not true at the last moment of time. As in the note above, I set this complicating factor aside. The argument of this chapter is that the *metric* version of Will Excluded Middle (Fn*p* $\vee$ Fn~*p*) fails, *even if* time is assumed to continue at least n units of time hence. *This* is the fundamental question at issue.

conjuncts with respect to conjunction. Similarly, we may face a decision whether to write *I gave money to Jones* or instead *I gave Jones money*—but this decision must be solely aesthetic or stylistic. Likewise for the items in (13) and (14). In this sense: *will* is scopeless with respect to negation.

Before moving on, it is important to note that MacFarlane is here plainly assuming that time does not end prior to tomorrow; in other words, his claim is that (13) and (14) say the same thing, under the assumption that time does not end prior to tomorrow. Clearly, (13) and (14) may come apart, if one is assuming that time ends a few minutes from now, well before "tomorrow". If time ends in a few minutes, (13) is presumably true and (14) false. The intuitive idea, again, is that (13) and (14) are equivalent in meaning, under the assumption that time does not end prior to tomorrow. In what follows, I make the relevant parallel assumption, and ignore this complication. It is worth noting, however, that this observation does complicate MacFarlane's contention that there is no important distinction in scope between (13) and (14).[7]

I deny that *will* is scopeless with respect to negation. In particular, I deny that (13) implies (14). The position I wish to defend in this chapter, *inter alia*, is that (14) implies (13), but (13) does not imply (14). Accordingly, these claims do not have the same meaning. Now, there is a sense in which I agree with these philosophers: it is, in ordinary contexts, extremely difficult to hear any distinction between (13) and (14)—that is, between ~Fn*p* and Fn~*p*. In ordinary contexts, that is, if you deny that the future features *p* in n units of time, then you affirm that the future features ~*p* in n units of time. This is because, in ordinary contexts, it is presupposed that there exists what we might call "the actual future"—and so, if it doesn't feature *p* in n units of time, it instead features ~*p* in n units of time. However, my contention is this. Once we move to (admittedly) non-ordinary, metaphysically-loaded contexts, in which we are explicitly considering the metaphysical model of the "open future", we can see that scopelessness breaks down. In such a context, there is, I argue, no reason to maintain that if it is not the case that the future features *p* in n units of time, it therefore follows that it instead features ~*p* in n units of time. Of course not: there is no such thing as "the actual future"! Contra the above authors, then, we cannot simply push the negation in ~Fn*p* inside the scope of the "F" to achieve Fn~*p*. More particularly, the claim that you can requires (or perhaps just *is*) a substantive *theory* of the future—a theory of the future that may indeed be plausible (and is certainly widespread), but a theory of the future that we could (I contend) nevertheless coherently reject. The inference holds, if it holds at all, as a matter of *metaphysics*, not semantics.

[7] Cf. Correia and Rosenkranz: "Are we therefore bound to conclude that if all future contingents fail to be true, they likewise fail to be false? Not obviously so. The principles $\neg F\varphi \rightarrow F\neg\varphi$ and $\neg F_n \varphi \rightarrow F_n \neg\varphi$ are objectionable on other grounds. They in effect rule out that time has come to an end: if time has come to an end, then $\neg F\varphi$, $\neg F\neg\varphi$, $\neg F_n \varphi$ and $\neg F_n \neg\varphi$ should all hold" (2018: 102). For a similar point, see Briggs and Forbes 2012: 12.

In order to see this result, we must again put on the table our three different models of the undetermined future, articulated in the previous chapter: (I) *Indeterminism with a privileged branch*, (II) *Indeterminism with an actual branch, but it is indeterminate which branch that is*, and (III) *Indeterminism with no actual branch*. And my claim is simple: because the third model has no actual future history, it invalidates the inference from ~Fn*p* to Fn~*p*. However, because the first two models do have (albeit in different ways) an actual future history, those models validate that inference. And the model implicit in ordinary discourse is the first model, and this is what makes it difficult to hear a distinction between the two claims. Nevertheless, the third model is perfectly metaphysically coherent—and this is the model of the "open future" I wish to defend. And my claim is that we are not in position to rule out the third model in virtue of semantic competence with *will* and negation.

To quickly recap. In the context of causal indeterminism, we have various "branches" that represent causally possible (maximal) ways things might go from here, consistently with the past and the laws. (Such branches will be segments of traditional abstract possible worlds.) On the model I defend, no one of these branches is metaphysically privileged, in the sense that that branch is uniquely "going to be". In this sense: there is no privileged branch in the model. Further, there is, now, no such thing as "the actual future history" or "the *actual* way things will go". (Thus, there is nothing, now, that deserves the title "the actual world".[8]) Here it is crucial to see that this model of the "open future" is to be distinguished from a different model of the open future, viz. model (II). This is the model of the "open future" presupposed by Cariani and Santorio (who are in turn inspired by Barnes and Cameron 2009); indeed, they maintain that their semantics

> presupposes that there is a 'unique' actual course of history. At the same time, it might be indeterminate which possible world instantiates the actual course of history. As a result, it might be indeterminate which world *will* selects, and *will*-statements may have indeterminate truth values. (2018: 131)

The (certainly mysterious) idea here is that though one such branch *is* the actual future history, it is just indeterminate which branch that is. On the model I defend, however, if this is what their semantics presupposes, then their semantics presupposes something false: it isn't "indeterminate" which branch *is* the actual future history—it is just that, again, there *is no* "actual future history" in the first place.[9] My aim in this chapter, however, is not primarily to compare this

[8] Cf. Kodaj 2013.

[9] Here we encounter a question for Cariani and Santorio: why is it safe to presuppose, when giving a semantics for *will*, that there is a unique actual course of history? And what happens to *will* claims when this assumption is denied? Perhaps their idea is that *everyone* agrees that there is such a unique actual course of history—the disagreement just concerning whether it is *determinate* what it is or

model of the open future to the former model. It is to assess the implications of the former model—which is, I believe, perfectly coherent.

But now we must consider the first model, viz., *indeterminism with a privileged branch*—a model that is often called "Ockhamism". According to Ockhamism, though the world is (or certainly may be) causally indeterministic, there are nevertheless facts about how causally indeterministic processes will unfold. To be sure, these facts are (standardly) humanly unknowable, but the facts are there nevertheless, and they are perfectly determinate.[10] To give expression to this model, various theorists have employed the idea of a privileged branch.[11] Now, my claim is simple: it is the Ockhamist's model that is implicit in ordinary discourse. In ordinary discourse, we might grant that there are various ways things might go from here. Nevertheless, we take it for granted that we can, *inter alia*, reason about, talk about, and bet about the facts concerning how things *will* go from here. Of course, a philosopher may come along and challenge our assumption that there *are* such facts. But now we are in a philosophical context—and even if we grant this philosopher his or her point, we may soon find ourselves saying things that would seem to belie it. That is: we have lapsed back into the ordinary context.

Indeed, that the Ockhamist model is the model presupposed in ordinary discourse is made plausible by the observation that, as soon as we adopt one of the other two models, we find it immediately difficult to give a philosophical account of our future directed talk.[12] Defenders of the other models may certainly try to address such worries, but the point is that they must be addressed. (See Chapters 6–8.) Ockhamism, however, generates no such difficulties. Instead, its difficulties are those of the metaphysician—namely, that it seemingly postulates a realm of fact that outstrips what could be accounted for by current physical reality and the laws alone.[13] And some feel that the existence of this realm of fact is objectionable. (See Chapter 1.) But the problems for the Ockhamist, then, are primarily metaphysical, not semantic.

indeterminate what it is. But this is false: certainly one standard way to express "openness" is simply to deny that there is a "unique actual future" at all (cf. Halpin 1988: 208–209, Belnap and Green 1994 (in Belnap et al. 2001: 133–136)), Hare 2011: 193, Pooley 2013: 340, Müller et al. 2019: 4, and Rumberg 2020: 357). One further note: Cariani and Santorio say that their semantics *presupposes* that there is a unique actual course of history. If Will Excluded Middle is meant to *follow from* their semantics, which simply *assumes* that there is such a unique actual course of history, then I have no objection, for, as I note shortly, such an *assumption* plainly validates Will Excluded Middle. However, Cariani and Santorio—and certainly MacFarlane—appear to write as if scopelessness should hold *no matter* our model—or that it is, in some sense, a semantic constraint on the coherence of such models. And it is this that I wish to deny. The assumption that there is such a unique actual course of history is explanatorily prior to the intuition that *will* is scopeless.

[10] For a defense of Ockhamism thus understood, see Rosenkranz 2012.

[11] Cf. Malpass and Wawer 2012, Wawer 2014, and Wawer and Malpass 2020.

[12] Cf. MacFarlane 2014: 233–236, and Williams ms., on, e.g., the "credence problem" and the "assertion problem".

[13] Cf. Belnap and Green (in Belnap et al. 2001: 168).

And now we can note the following. The Ockhamist's model is plainly a model on which the distinction between ~Fn*p* and Fn~*p* is simply unimportant. Intuitively: if there are (determinate, fully complete, fully exhaustive) facts about how indeterministic processes will unfold, then if those facts don't have it that *p* in n units of time, then they have it that ~*p* in n units of time. That is, we might say, just in the nature of "the facts". Slightly more carefully, in terms of the model, we might notice that it immediately vindicates the following pattern of reasoning:

(1) It is not the case that: the actual future history features *p* in n units of time. (~Fn*p*)
(2) There is an actual future history, which, for any *p*, either includes or excludes *p* in n units of time.
(3) So, the actual future history features ~*p* in n units of time. (Fn~*p*)

And there we have it: models with a unique actual future immediately validate the inference from ~Fn*p* to Fn~*p*. Further, such a model is, very plausibly, the model implicit in ordinary discourse. More particularly: I contend that, in ordinary discourse, the second premise is simply *implicit* and *unspoken*. This claim is, in some sense, simply a regulative principle undergirding ordinary thought and talk about the future. And this opens up the space to maintain the following: if a revisionist metaphysician *denies* the second premise, then that metaphysician will likewise have reason to deny the inference from ~Fn*p* to Fn~*p*. And if that metaphysician's *model* is itself coherent, then the inference will, indeed, by licensed by *certain* models of the future—but it is not one that should be licensed by semantic competence alone. More generally: I maintain that it is *because* we already implicitly accept premise (2) that we feel like we can move from ~Fn*p* to Fn~*p*. Thus: proponents of models (I) and (II) cannot appeal to linguistic data to the effect that we do not recognize a distinction between ~Fn*p* and Fn~*p* to *support* premise (2)—for that data *presupposes* premise (2).

3.2 Neg-Raising: A Primer

The way I wish to develop this point is to develop the claim that *will* is a so-called *neg-raising* predicate. (I extend this explanation to the counterfactual *would* in the next chapter.) "Neg-raising" refers to the widespread semantic phenomenon whereby what is in fact semantically wide-scope negation gets treated, in context, as if it belonged to the relevant embedded clause. For instance, *I don't think that Trump is a good president* strongly tends to implicate *I think that Trump is not a good president*—despite the former not semantically entailing the latter. It is tempting simply to let Laurence Horn—whose work on negation is as comprehensive as it is careful—explain the phenomenon in his own words. And so I will. Thus Horn (writing here with Wansing):

> In his dictum, "The essence of formal negation is to invest the contrary with the character of the contradictory", Bosanquet encapsulates the widespread tendency for formal contradictory (wide-scope) negation to be semantically or pragmatically strengthened to a contrary... The strengthening of a contradictory negation... to a contrary typically instantiates the inference schema of disjunctive syllogism or *modus tollendo ponens* in (11):
> (11) A ∨ B; ~A; B

Note: A and B are here *contraries*: both can be false, but not both true. However, given the first premise, B, in effect, becomes the contradictory of A; that is, if we have ~A, we have B, and vice versa. Horn and Wansing continue:

> While the key disjunctive premise is typically suppressed, the role of disjunctive syllogism can be detected in a variety of strengthening shifts in natural language where the disjunction in question is pragmatically presupposed in relevant contexts.
>
> Among the illustrations of this pattern [is the] tendency for negation outside the scope of (certain) negated propositional attitude predicates (e.g. *a does not believe that p*) to be interpreted as associated with the embedded clause (e.g. *a believes that ~p*); this is so-called "neg-raising".
>
> When there are only two alternatives in a given context, as in the case of neg-raising (as stressed by Bartsch 1973; cf. Horn 1978; Horn 1989, Chapter 5), the denial of one... amounts to the assertion of the other. The relevant reasoning is an instance of the disjunctive syllogism pattern in (11), as seen in (12), where *F* represents a propositional attitude and *a* the subject of that attitude.
>
> (12) *F(a,p)* ∨ *F(a, ~p)* [the pragmatically assumed disjunction]
>
> *~F(a,p)* [the sentence explicitly uttered]
>
> *F(a,~p)* [the stronger negative proposition conveyed]
>
> The key step is the pragmatically licensed disjunction of contraries [*a believes that p* ∨ *a believes that ~p*]: if you assume I've made up my mind about the truth value of a given proposition *p*, rather than being ignorant or undecided about it, then you will infer that I believe either *p* or *~p*, and my denial that I believe the former will lead you to conclude that I believe the latter.[14]

[14] Horn and Wansing 2017. Almost identical points can be found in Horn 2015 (and the classic Horn 1989). However, in the interest of simplicity, I have followed the more streamlined presentation of these points in Horn and Wansing 2017. Interestingly, elsewhere in their 2017, Horn and Wansing mention "future contingents" as a case in which one might reasonably claim that propositions often

The key idea here is this: *a believes that p* and *a believes that ~p* are strictly speaking contraries: both could be false. However, if the disjunction of these contraries is presupposed in context, then *a does not believe that p*—what is strictly speaking [~(*a believes that p*)]—will tend to be interpreted as *a believes that ~p*. That is, what is in fact semantically wide-scope negation gets interpreted as if it were associated with the "embedded clause". As Horn notes, however, neg-raising effects are not witnessed solely in cases of (certain) propositional attitudes. Commenting on Horn, Gajewski (2007: 292) summarizes:

> *A list of Neg-Raising predicates, arranged by semantic field (Horn 1989):*
> a. think, believe, suppose, imagine, expect, reckon, feel
> b. seem, appear, look like, sound like, feel like
> c. be probable, be likely, figure to
> d. want, intend, choose, plan
> e. be supposed to, ought, should, be desirable, advise, suggest

The central idea is that in all of these cases, there is a semantic difference between the relevant wide-scope and narrow-scope readings (*I don't think that p/I think that ~p; I don't want to do it/I want to not do it; You're not supposed to do that/You're supposed to not do that*). However, *in context*, this distinction is often suppressed or otherwise masked—and this is because, in these contexts, we bring a certain *model* of the situation with us. In a context in which I am assuming that you're not simply indifferent (that you aren't indifferent is part of my background model of the situation), when you say that you don't want to come to the party, I hear this as an assertion that you want to *not* come. Indeed, it is extremely difficult to hear *I don't want to come* as anything *but* this stronger assertion—or, perhaps, it is difficult to imagine that there even *is* a stronger assertion available (*He said he doesn't want to come! Quit asking him.*). At the same time, on reflection, we are capable of seeing that it is *possible* (even if, in context, probably unkind) for someone truly to say *I don't want to come*, although that person doesn't want to *not* come—because, at the moment, that person is completely indifferent. That is, on reflection, we can grant that *does not want to come* does not *semantically* entail *wants not to come*. It does so only *holding fixed* our assumed model of the situation, viz., that you aren't indifferent. If, indeed, you *aren't* indifferent, then that you don't want to come does imply that you want not to. But once this model is relinquished, a scope distinction becomes salient that was otherwise practically irrelevant.

taken to contradictories are in fact contraries: "Other cases in which apparent contradictories can be seen as contraries, and thus immune from any application of LEM, are future contingents (*There will be/will not be a sea battle tomorrow*)." The authors do not, however, link this issue to the issue of neg-raising.

And this is plainly deeply similar to what I wish to say about *will* and *will not*. Indeed, my claim is that we may plausibly add *will* to the list of neg-raising predicates above.[15] In this case, the relevant inference pattern identified by Horn, I contend, goes as follows:

(1) There is a "a unique actual course of history", which, for any *p*, either includes or excludes *p* in n units of time. (Implicit, unspoken assumption)
(2) F*np* ∨ Fn~*p*. (Trivial consequence of (1)—and establishment of semantic contraries) (e.g. There will be a sea-battle tomorrow or there will not be a sea-battle tomorrow)
(3) ~Fn*p* (The proposition considered or uttered) ("It is not the case that there will be a sea-battle tomorrow"/"I deny [assert that it is false] that there will be a sea-battle tomorrow")
(4) Fn~*p* (The proposition communicated or expressed) ("There will not be a sea-battle tomorrow")

My argument, then, is that we can see the distinction between MacFarlane's original items in (13) and (14) as continuous with a wide range of linguistic data in which distinctions of scope are suppressed or masked by the implicit models we bring to the relevant contexts. At the same time, the difference between

[15] In this chapter, I defend the claim that will is a "neg-raiser". However, some might have a conception of "neg-raising" that prohibits *will* from being a neg-raiser from the start. Roughly, on a certain *syntactic* approach to neg-raising (cf. Collins and Postal 2014), neg-raising is essentially a bi-clausal operation (in which the negation is "raised" from the lower clause). In personal correspondence, Laurence Horn thus notes that if neg-raising is essentially such an operation, if *will* is a modal, *will* cannot be a neg-raiser, since a modal is in the same clause as the main verb it governs. However, some in the literature have employed a more permissive approach to the terminology of "neg-raising"; as just noted by Gajewski, for instance, Horn (1978: 198) claims that the deontic modal *should* is a neg-raiser, on grounds that *I don't think he should go to the party* strongly conveys *I think he should not go to the party*. In this book, I thus assume a conception of "neg-raising" that does not prohibit *will* from being a "neg-raiser" on terminological grounds alone; in other words, I assume a conception on which *should* is appropriately called a "neg-raiser" (and thus, in principle, a conception on which *will* can be a neg-raiser). For more on these issues, see Collins and Postal 2017, which clarifies the relationship between the syntactic theory of neg-raising developed in their 2014 and the pragmatic/excluded middle approach to which I appeal in this chapter. Collins and Postal 2014 reserve the label "Classical Neg-Raising" (CNR) for neg-raising in the more narrow sense just noted; I thus assume that not all neg-raising is classical neg-raising.

Incidentally, the comparison with *should* is instructive. (Cf. Horn's (1978: 200) discussion of St. Anselm on the Latin *ducere*.) "Trump should be impeached or Trump should not be impeached" certainly *sounds* initially like an instance of ($p \vee \sim p$), although, on reflection, we may be prepared to grant that it isn't; for someone on the fence, it is not the case that he *should* be impeached, and not the case that he should *not* be impeached. So similarly, when reality is "on the fence" concerning Trump's impeachment tomorrow, I say, it is not the case that he *will* be impeached tomorrow, and not the case that he will *not* be impeached tomorrow. It is open. (Thanks to Laurence Horn for discussion on these points.) I extend this comparison with *should* towards the end of the next chapter. (See further Todd and Rabern forthcoming.) A preview: "No one nowadays should have to suffer from malaria" certainly seems *prima facie* equivalent to "Everyone nowadays should be free from malaria", but very plausibly we cannot rely on this feeling of equivalence to generate an argument for "Should Excluded Middle"!

the case of *will* and *will not* and the standard cases of neg-raising listed above is perhaps obvious. For instance, as explained, in context, we may bring with us a supposition that Jones has thought about the matter and formed an opinion one way or the other; when Jones thus says that he doesn't think that *p*, we interpret this to mean that he thinks that ~*p*. But it is, for most of us, relatively easy to forgo or cancel this assumption; all of us are familiar with the situation of withholding belief (or being agnostic). Thus, when such an option is made salient, we are able to consider *Jones does not think that p* as true and *Jones thinks that ~p* as false. He's an agnostic on this matter, so the inference that usually holds good does not hold good.

In this case, then, the situation that makes salient the difference in scope is relatively familiar and benign. But this plainly is not the case for *not that will* and *will not*. Indeed, the purported situation that would make salient the difference in scope is a situation that would only be insisted on by a philosopher. *Only a philosopher*—a philosopher!—would think to question the inference from (3) to (4), because only a philosopher would have cause to consider and reject (1) (and thereby (2)). Only philosophers (broadly conceived) are concerned with "models of the undetermined future", and only philosophers would contend that, as far as we know, *indeterminism with no actual future branch* is the correct such model. Ordinary persons in ordinary life may pause to say *Well, wait—maybe Jones hasn't considered the matter, so, sure, he doesn't think that p, but maybe he also doesn't think that ~p*. But only a philosopher would wish to pause to say, *Well, wait—maybe there are no facts about undetermined aspects of the future, so, sure, it isn't yet the case that it* will *happen an hour from now, but maybe it also isn't yet the case that it* won't *happen an hour from now*. This, then, plausibly explains why *will* has not (to my knowledge) yet appeared on any list of neg-raising predicates. Semanticists and linguists concerned with the theory of neg-raising are plainly not going to be concerned with cases that appear, if at all, only in the context of an explicit rejection of a metaphysical theory of time. (Their work is already difficult enough.)

And so the difference is this. The situation, I contend, that masks the relevant scope difference in the case of *will* is *thoroughgoing* and *metaphysically entrenched*. However, the claim that the scope differences are *there* is deeply theoretically motivated. My claim, then, is that the move from ~Fn*p* to Fn~*p* is, in fact, a move not licensed *merely* by semantic competence, but semantic-cum-*metaphysical* competence—that is, competence with the prevailing *metaphysical theory* of the future and what semantical distinctions that theory makes relevant and irrelevant. On reflection, that is, there is a coherent (albeit highly controversial) metaphysics that makes salient the given distinction in scope. I do not know what it is that entitles us to accept premise (1) above, if indeed anything does, but what I *do* know is that, *if* something does, it isn't linguistic or semantic competence with *will* and negation. After all, it is *because* that (1) that (2)—and because that

(2) that we can move from (3) to (4). That we can move from (3) to (4) *presupposes* exactly what is at issue: that there is some future branch which is the unique actual future.

3.3 Interlude: Pure Semantic Competence

Here we must make an important clarification. My claim is that we must be careful to distinguish between judgments made in virtue of *pure* semantic competence, and those made in virtue of semantic-cum-metaphysical competence. However, my view is *not* that our faculties of semantic competence are simply broken, and *any* semantic distinction can be introduced by inventing some bizarre metaphysics which allegedly brings it out. For instance, I suggest that it is in virtue of *pure* semantic competence that we can distinguish between

John has yet arrived to the party
John has not yet arrived to the party.

More particularly, it is in virtue of pure semantic competence that we judge that the former sentence is infelicitous, and the latter felicitous. ("Yet" is a so-called negative polarity item, hence the infelicity of the former.) But it is extremely difficult to see what kind of *metaphysical* hypothesis might imply that former is, after all, felicitous. Similarly, it is pure semantic competence that tells us that there is no difference in meaning between

Sally gave money to John
Sally gave John money

Then again: perhaps some theory of substances and relations may make one true and the other false? Well, having considered the matter for a few minutes—no, I don't think so, but never mind. Needless to say, it is beyond the scope of this book to attempt to give some sort of criterion that might distinguish between judgments of pure semantic competence and the rest—if, as I doubt, such a criterion could be given at all. The important point is this. Model (III) cannot be ruled out on *merely* grounds of semantic competence.

3.4 Against Scopelessness: Quantifiers

Recall Cariani and Santorio's claim that *will* is scopeless with respect to "an important class of other linguistic items". Thus far, the focus of this chapter has been on whether *will* is scopeless with respect to *negation*. Here I focus on whether *will* is scopeless with respect to quantifiers.[16] Notably, on this issue, Jonathan Bennett—who is no friend of *counterfactual* excluded middle (about which more in the next chapter)—sees no distinction between "Will ∃x Fx" and "∃x Will Fx". Indeed, Bennett writes:

> In the case of Will [unlike with *Would*, Bennett maintains], there is no room for two readings, no issue about the scope of the quantifier... It cannot be that he will appoint a woman unless there is a woman whom he will appoint.[17]

I will argue, however, that Bennett was mistaken on this point.

Suppose we are running an indeterministic lottery in one hour with three and only three tickets. Now consider the sentence:

(1) No ticket will win (in an hour).

As a first approximation, it seems that someone uttering (1) must be maintaining that our lottery is going to be a failed lottery—a lottery in which no ticket has ended up winning. In other words, on the assumption that any non-win is a loss, (1) seems equivalent to "Every ticket will lose." However, (1) is *de re/de dicto* ambiguous (respectively):

(1a) No ticket is such that it will win (in an hour). ~∃x, x a ticket, x is such that: **Will**n: x a winner

(1b) It will be (in an hour) that no ticket wins. **Will**n: ~∃x, x a ticket, x a winner

Now, everyone can observe the *syntactic* distinction between (1a) and (1b). A theory on which *will* is "scopeless", however, predicts that (1a) and (1b) are nevertheless truth-conditionally equivalent. *Prima facie*, this seems reasonable. If no ticket is such that it will win at a certain time, isn't this just to say that it will be that no ticket wins at that time?

It is not. I contend that, if the future is open, there are scenarios in which (1a) is true and yet (1b) is false. On reflection, if there are no facts about undetermined aspects of the future, then such a scenario is easy to construct. Suppose that there are three and only three total, distinct branches. On branch 1, ticket 1 wins, on branch 2, ticket 2 wins, and on branch 3, ticket 3 wins. Now, since there are no facts about undetermined aspects of the future, no one of these tickets is such that *it* will win. If one such ticket *were* such that *it* will win, then this very fact—that

[16] Cf. Forbes 1996 for discussion of related issues; Forbes does not, however, consider the position I develop shortly. Cf. also Higginbotham 1986, who introduced the method of using quantifiers to test the scope of negation for conditionals.

[17] Bennett 2003: 186.

that ticket will be the winner—would be a fact about an undetermined aspect of the future. (1a), accordingly, is true. However, we have also said that our three branches are all and only the branches there are. And observe: on every branch, some ticket or other is the winner. On branch 1, it is ticket 1, on branch 2, ticket 2, and on branch 3, ticket 3. Thus: on every branch, in n units of time, we have the formula '∃x, x a ticket, x a winner'. What we have is simply different tickets making that formula true on the different respective branches. But the formula is true on *all* the branches. Thus, far from having (1b), we in fact have the following: **Willn:** ∃x, x a ticket, x a winner. (1a) is true in this scenario and yet (1b) is false.[18]

I am prepared for the obvious objection: "Are you really saying that there is a true reading of 'No ticket will win' in a context in which it is taken for granted that it is *open* that ticket 1 wins, open that ticket 2 wins, and open that ticket 3 wins?" And yes, that is what I am saying. I contend that the default reading of (1) is (1b), however, and that in this scenario, (1b) is false. (1b), of course, is equivalent to the claim that every ticket will lose. You are, therefore, ill-advised blithely to assert (1) in such a context. Nevertheless, (1) is ambiguous between (1a) and (1b), and (1a) is true in this scenario. (If you want to say that "No ticket will win", because, on your view, the future is open, and so no ticket is such that *it* will win, then, even though what you say is true, you had better prepare your audience.) That (1) has a true reading in this scenario is perhaps unexpected, but there it is. As I see it, however, you can earn the right to make fun of this view only if you can earn the right to make fun of the view that there are no facts about undetermined aspects of the future. For if there are no facts about undetermined aspects of the future, and we are running an indeterministic lottery with three tickets, it is just plain true—or so it seems to me—that no one ticket in this lottery is such that *it* will win. And yet if some ticket or other wins on *every* way things could develop for this lottery, it is just plain false that it will be that no ticket wins. Indeed, it is just plain true that it will be that some ticket wins. It will be that some ticket wins, but no one ticket is such that *it* will win. This is strange, but the open future is strange. I don't know what else to tell you.

The general lesson is this. We cannot unproblematically move between the following:

Willn: ∃x ϕx (e.g., "It will be that there is a winning ticket in an hour")
∃x **Willn:** ϕx (e.g., "There is a ticket such that it will be the winning ticket in an hour.")[19]

[18] This issue here is a variant on the following. On my view, we could have ~Fn*p*, ~Fn*q*, ~Fn*r*, and yet Fn($p \vee q \vee r$). For suppose we have three and only three differing branches: a *p* branch, a *q* branch, and an *r* branch. Then, on every branch, in n units of time, we have it that ($p \vee q \vee r$). After all, any *p* branch is ipso facto a ($p \vee q \vee r$) branch, and any *q* branch is ipso facto a ($p \vee q \vee r$) branch, and so on. Thus, in this case, Fn($p \vee q \vee r$) is not a future contingent, but a future *necessity*.

[19] Note: there is an important subtlety here that I am ignoring for the sake of simplicity. There is of course an important sense in which *any* presentist must deny the equivalence of these formulas. For instance: a presentist may want to grant that it will be in 1000 years that there exists a Martian outpost.

I appreciate that these points are delicate. Yet it is worth bringing out the way in which they seem problematic for scopelessness. Scopelessness predicts that there is *no difference in meaning* between '**Will** $\exists x\ \phi x$' and '$\exists x$ **Will** ϕx'. Accordingly, whatever semantic profile is had by one ought to be had by the other. There are two distinct challenges here. The first—challenge (a)—is simply the challenge that there are scenarios in which one such claim seems more clearly true than the other. Again, in the three branch scenario in which, in branch 1, ticket 1 wins, and branch 2, ticket 2 wins, and branch 3, ticket 3 wins, the following claim is clearly true (indeed, in MacFarlane's terminology, it is settled-true): **Willn**: $\exists x$, x a ticket, x a winner. But whereas this claim seems, for this reason, clearly true in this scenario, the other given claim seems much less clearly true. In this scenario, it is not clearly true that *one* of these tickets is such that *it* will be the winner (in n units of time). (At least, the question whether this claim is true is a confusing question.) The second challenge—challenge (b)—is related to the first. For, again, *this* claim (the *de re* claim) would seem to be committing us to facts about the future that outstrip what is determined by the present; not so, of course, for the former claim, which is *settled* by the present. For whichever ticket has the property in question—and, on this view, some ticket has it—its having this property cannot be accounted for in terms of facts about current conditions and causal laws, and its having it will therefore be a brute fact with respect to such conditions and laws. Insofar as one such claim seems clearly true whereas the other does not, and insofar as one such claim seems to have ontological and metaphysical commitments that the other does not, we have reason to think that these claims, contra scopelessness, do not mean the same thing.

But perhaps we are now in position to see how scopelessness might be saved. Concerning challenge (a), the proponent of scopelessness must say the following. In this scenario, it is indeed clearly true that one of these tickets is such that it will be the winner. It is just *indeterminate which* is such that it will be the winner. Again, some ticket or other has the property *being going to be the winner*. It is just indeterminate which ticket has that property. In this way, we can preserve the

But our presentist will presumably *not* wish to grant that there exists something such that it in 1000 years will be a Martian outpost (what would *that* be?)—anyway, not unless our presentist is prepared to say that the domain of objects never changes over time (nothing really comes into existence or goes out of existence). Notably, however, an eternalist does not have this reason for distinguishing between these two formulas; for her, if there will be a Martian outpost in 1000 years, there indeed exists something such that it in 1000 years will be a Martian outpost. For the eternalist, our most unrestricted quantifier *does* range over future objects like Martian outposts (if there will be such outposts). For discussion of this issue, see Prior 1957: Ch. 4, and Sider 2006. I want to sidestep this issue simply by restricting our attention to contexts in which the objects in the given domain exist, even on presentist grounds. In other words, *all* presentists will have cause to say that "It will be that there exists a winning ticket" does not entail "There exists a ticket such that it will be the winning ticket" if, say, the lottery is fated to be run next year, and there aren't even *tickets* for the lottery yet. Nevertheless, if all the tickets for the lottery have just been created, then a presentist who accepts scopelessness will insist that there is no difference between "It will be that there exists a winning ticket" and "There exists a ticket such that it will be the winning ticket". And *that* is what I am denying.

equivalence between the relevant formulas. If it will be that some ticket wins, then, indeed, some ticket is such that it will win. But if it is indeterminate which ticket will win—look at the disagreeing branches!—then it is indeterminate, now, which ticket is such that it will win, although, of course, some one ticket is indeed such that it will win. Again: there is a ticket such that it will win, but it is indeterminate which ticket has this property.

So far so good. This is an elegant reply to problem (a). But is it an adequate reply to problem (b)? Problem (b) is that '∃x, x a ticket, x is such that: **Willn:** x a winner' seems to commit its proponent to a fact that goes beyond what is entailed by the present, whereas '**Willn:** ∃x, x a ticket, x a winner' does not. And here the proponent of scopelessness arguably must say the following. Yes, there is indeed a fact about the future that goes beyond what is entailed by the present and the laws. This is precisely the fact about which ticket is such that it will win—for there is indeed a particular ticket such that it will win. This fact—about which ticket has this property, and again, some ticket does have this property!—cannot be accounted for by virtue of the present and the laws. The central claim the proponent of scopelessness must make here is that there are facts about the future that go beyond what is entailed by the present and the laws—there are *primitive future directed facts* (recall Chapter 2)—but it is simply *indeterminate* what these facts are.

But now I wish to harken back to Chapter 1 and the motivations for the theory of the "open future" at issue. As I am imagining it, precisely the motivation of the open futurist is that there are *no facts about the future that go beyond those that are entailed by the present and the laws.* Their central motivation is *not* thereby respected if we say that though there are facts about the future that go beyond the present and the laws, it is indeterminate what those facts are. No! That would be precisely to violate the spirit of the motivation for openness, which is the feeling that *any such facts*—say that they are "indeterminate" if you must—would be mysteriously brute or unexplained. Consider once again our comparison with fictions. There is all the difference between saying that there are no facts about fictions that go beyond those specified by the fiction-determining facts, and that though there are facts about fictions that do go beyond those specified by the fiction-determining facts, it is simply indeterminate what facts those are. On the latter view, but not the former, we still must ask: where are those facts *coming from*?

Perhaps we are at some kind of stalemate. Perhaps we have simply discovered two distinct, perfectly respectable conceptions of "openness", one which would preserve scopelessness, and one which would not. And so let me simply put my cards on the table. My claim is that we do not *need* to have recourse to the claim that (in this scenario) there is a ticket such that it will be the winner, but it is indeterminate which ticket that is. We can simply say that though it will be that a ticket wins, there is, as of now, no ticket such that *it* will win. Here we simply employ the familiar tools and distinctions we already had at our disposal, without introducing a new, mysterious kind of "indeterminacy". The cost, of course, is that we now must defend some scope distinctions to which we previously had not been sensitive. But this cost is well worth paying. As shown in the neg-raising literature, it is often the case that a scope distinction is not salient until a shift in one's background assumptions makes it salient. My contention is that an acceptance of the claim that there are no facts about undetermined aspects of the future is precisely a claim that makes salient the scope distinctions developed above.

3.5 One or the Other/Neither

We are not out of the woods yet. There are other *de re/de dicto* ambiguities to contend with. Consider an indeterministic lottery with two and only two tickets. There are two and only two branches; in n units of time, on branch 1, ticket 1 wins, and on branch 2, ticket 2 wins. Now suppose someone says:

(3) One or the other will win.

As before, I think that (3) has a true reading in this scenario, and a false reading—to be sure, a false reading that must be forced, but a reading that can be forced. The true reading again is the *de dicto* reading (3b), and the false reading is the *de re* reading (3a):

(3a) One ticket is such that it will win or the other ticket is such that it will win.
(3b) It will be that one ticket wins or the other ticket wins.

In this scenario, I contend that (3a) is false, whereas (3b) is true.

A final case. Scenario: an indeterministic three ticket lottery; I buy tickets 1 and 2, Jones buys ticket 3. There are three branches as before; on branch 1, ticket 1 wins, etc. Jones is prone to superstition; in a superstitious moment, he looks at my tickets and says:

(4) Neither will win.

Do I agree with what Jones said? That depends. There are again two readings of (4), the *de re* (4a) and the *de dicto* (4b):

(4a) Not either are such that they will win.
(4b) It will be that not either win.
= both will lose.

If Jones means (4a), then I agree: neither of my tickets are such that they will win. That sounds bad for me. However, I do *not* thereby think that both of my tickets are such that they will *lose*. Under one reading of (4), I agree, and on the other, I disagree. Neither of my tickets are such that they will win, but happily neither are such that they will lose. That's why I want to play this lottery.

3.6 A Prediction of Salience

If you are, at this stage, at least somewhat bewildered, then I am sympathetic. For though I claim that the given scope distinctions are *there*, they certainly do not *feel* like they are there, and we certainly do not proceed practically as if they are there. And this I am happy to grant. But I would like to defend these scope distinctions by making a certain kind of prediction of my own—a prediction about what we should expect to see *if* there came to be a community of speakers who were determined to speak only in ways licensed by the philosophical theory that there are no facts about undetermined aspects of the future. For it is important to note that ours is certainly not anything like such a community. What sorts of distinctions may *become* salient to those in such a community?

Here is an initial comparison. Previously we could reliably communicate to our interlocutor that we think Trump is not a good president simply by saying, in a particular tone and context, "No, I don't think Trump is a good president." But now 1000 neutrals have moved to town. Eventually, I can no longer reliably communicate to my interlocutor that I think Trump is not a good president by saying that I don't think Trump is a good president. For now my interlocutor may wonder: are you simply *denying* that you think he's a good president, which is consistent with you not thinking he is *not* a good president? In other words: are you possibly yet one more of the neutrals? A scope distinction has been made salient that otherwise was practically irrelevant.

Similarly: we've become open futurists who take very seriously our open futurism in our daily thought and talk. (Don't ask me why we've made this mistake.) Previously we could reliably communicate to our interlocutor that it is going to fail to rain tomorrow simply by asserting that it isn't going to rain tomorrow. But now we recognize that there are no facts about the future beyond those necessitated by the present—and people know that and talk accordingly. We're making critically important plans and wondering about the weather; you report that Jones—an authority about the weather—said it isn't going to rain tomorrow, but then the phone suddenly cut out. That's good news. He said it isn't going to rain. But wait. The phone cut out. So, sure. He said it isn't *going to* rain tomorrow. But did he say it is going to *not* rain tomorrow? Now we want to be clear. Did it sound like he might be about to say that though it isn't *going to* rain tomorrow, as of yet, it isn't going to *not* rain tomorrow either? In which case: pack umbrellas.

My prediction, in short, is that though this *sounds* odd to us, conversations like these would eventually encourage our (perhaps benighted) open futurists to hear certain scope distinctions as salient that we do not find salient. Would this simply be a scenario in which the relevant words—*will, be going to*—have taken on new meanings? For instance, one might object that, in the final line of the previous paragraph, the italicized *going to* simply means something different than our *going to*. In particular, one might object that, in this usage, it means (roughly) "determined". ("Did it sound like he might think that though it isn't determined for there to be rain, it isn't determined that there be no rain either?") Does *going to* simply mean "determined" in this scenario? No. It is well known that which "reading" of a given sentence we find salient in a context is highly sensitive to such factors as the tone, emphasis, and intonation of the speaker. The given emphasis makes a certain reading more salient or otherwise possible, but it does not change the *meaning* of the *going to*.

Consider. The following sentence almost inevitably gives rise to the "neg-raised" reading:

Jack: Jill, listen. I don't want to go to your party.
Jill: OK, well, maybe next time.
[Improper response: OK, let me try to sway you!]

But with a change in emphasis, the neg-raised reading is at least postponed:

Jack: Jill, listen. I don't *want* to go to your party.
Jill: OK, but... what? You are going to come anyway? Or you're on the fence? Or what?
Jack: But... right. I don't want to *not* come. Given my anxiety, I just feel very unsure about parties right now.

In this second dialogue, *want* has not become *definitely want*. It is, well, just *want*. It is just that a change in emphasis has made a reading salient that otherwise would not have been salient; the change in emphasis invites a "But...". Similarly, if we say that it isn't *going to* rain tomorrow, this invites the possibility of saying, "But it isn't going to *not* rain tomorrow either", and it can do so without involving a change of the meaning of *going to* to anything like *definitely/determinately/determined to be going to*.

3.7 The Dialectic: Circular Arguments

Given the above, consider the following argument—implicit in MacFarlane's argument above—against what I have called "model (III)".

(1) WEM is a logical truth (and *will* is "scopeless"). (Generalization from linguistic data)
(2) If WEM is a logical truth (and *will* is "scopeless"), there is a unique actual future (and future contingents aren't all false). So,
(3) There is a unique actual future (and future contingents aren't all false).

My reply is that the argument is objectionably circular. Premise (1) seems plausible (to the extent that it does) *because* we already think of the conclusion as true. The argument thus gets us nowhere. It is at least in part because we implicitly think that there is a unique specification of "what is actually going to happen tomorrow" (i.e., that there is a unique actual future) that it seems to us that WEM is a logical truth, and that *will* is "scopeless", and that therefore future contingents can't all be false. Thus, that it seems like WEM is a logical truth, and that *will* is scopeless, can't be *reason* to believe that there is a unique actual future, and that therefore future contingents can't all be false.

3.8 Some Comparisons with Other Modals

Cariani and Santorio maintain that *will* is a "modal". However, they claim, *will* simply has a unique property amongst modals: it is scopeless. At this stage, however, it is worth pausing to remark on precisely how (in Cariani and Santorio's words) "remarkable" scopelessness really would be, if *will* is a modal. (And even if *will* is not a modal, some comparisons are nice.) As they note, that *will* is scopeless would be a unique feature of *will* amongst other modals: to my knowledge, there are no *other* modals M—of whatever "flavor"—such that there is no truth-conditional difference between ~M*p* and M~*p*.[20] Consider:

[20] What counts as a "modal" in this context? It is difficult to say (cf. Pullum and Huddleston 2002: 172). At any rate, Cariani and Santorio contrast *will* with *must* and *might*, which, they say, are modals that do not commute freely with negation—and those I have listed here seem standard. For an updated version of the modal view defended by Cariani and Santorio, see Cariani (2021). I set aside the claim that *negation itself* is a modal (for discussion, see Berto and Restall 2019); negation would seem to commute freely with itself.

A further note: we can observe that there are indeed modals—possibility modals—in which ~M*p* does imply M~*p*, although not the other way around, e.g. metaphysical possibility. Under standard assumptions, ~ Possibly *p* implies Possibly ~*p*, though not vice versa. "Possibly" is thus not "scopeless", although ~ Possibly *p* does imply Possibly ~*p*. (Similar claims can be made concerning epistemic possibility and deontic permissibility.)

What is thus (minimally) required for the claim that ~Fn*p* does not imply Fn~*p* is the claim that *will* is stronger than a *possibility* modal. And, indeed, this is plausible: intuitively, to say that something *will* happen is to say something stronger than that it is *possible* for it to happen. Intuitively, in terms of modal strength, *will* lays between *may* and *must*: it is stronger than *it may happen* and weaker than (but of course does not *rule out*) *it must happen*.

Incidentally, we can connect this claim about *will* with a theme from Horn on the nature of neg-raising, although a full discussion of these issues must lie outside the scope of the present chapter. In his 1975, Horn considers as a necessary condition on neg-raising something he calls "midscalar

(i) Necessity. ~(*Necessarily p*) does not mean *Necessarily ~p*.
'*Necessarily p* ∨ *Necessarily ~p*' eliminates *contingency*.

(ii) Must (epistemic). ~(*According to S, it must be that p*) does not mean *According to S, it must be that ~p*.
'*According to S, it must be that p* ∨ *According to S, it must be that ~p*' eliminates *uncertainty*.

(iii) Obligation. ~(*S is obligated to see to it that p*) does not mean *S is obligated to see to it that ~p*.
'*S is obligated to see to it that p* ∨ *S is obligated to see to it that ~p*' eliminates *mere permission*.

(iv) Belief. ~(*S believes that p*) does not mean *S believes that ~p*.
'*S believes that p* ∨ *S believes that ~p*' eliminates *agnosticism*.

(v) Intention. ~(*S intends to bring it about that p*) does not mean *S intends to bring it about that ~p*.
'*S intends to bring it about that p* ∨ *S intends to bring it about that ~p*' eliminates *indecision*.

My point here is not that since scopelessness fails in these cases, *will* cannot be scopeless. Perhaps there is an important disanalogy between *will* and these other cases. My point, instead, is twofold. First, it is burden-shifting: proponents of scopelessness must explain what this disanalogy comes to. Second, and more importantly, it is illustrative: if it is claimed, from the outset, as a semantic "constraint" on our theorizing, that the items at issue in (i)—(v) are scopeless, then, from the outset, we seemingly eliminate as possibilities *contingency*, *uncertainty*, *mere permission*, *agnosticism*, and *indecision*. And the argument of this book is that if is it similarly insisted that *will* is scopeless, then we eliminate, from the outset, what I would like to call *openness*. My theory is that openness stands to *will* as mere permission and contingency stand to *obligation* and *necessity*.

If we defend the claim that *will* is a modal, we thus have a choice: we can defend the claim that *will* is unique amongst modals in being scopeless with respect to negation, or we can maintain that *will* is unique amongst modals in having standard scope interactions with negation, but these interactions being systematically suppressed in ordinary thought and talk by our implicit assumptions about its unique *subject matter*—namely, the future. The former approach must see *will*

generalization". (For discussion, see Gajewski 2005: 86–90.) Roughly, the idea is that expressions in the same semantic field can be ordered in terms of logical strength, e.g., *some, many, most, all*—and that, in order to be a neg-raiser, the expression must be somewhere in the middle. (For more on this theme, see Pullum and Huddleston's (2002: 838–843) discussion of "increased specificity of negation" in terms of "medium strength modality".) My point here is not that neg-raisers must be "medium strength" in the requisite way; my point is instead that, if it is claimed that they must be, then one could plausibly contend that *will* is "medium strength".

as semantically discontinuous with other modals. The latter approach instead can see *will* as perfectly semantically continuous with other such modals—and, indeed, can see the suppression of the relevant scope distinctions as perfectly continuous with a whole range of *distinct* semantic data (identified in the neg-raising literature) in which we can observe precisely the phenomenon I have here identified. The latter option sees deep continuity where the former sees discontinuity. The latter option, to this extent, is preferable.

3.9 Some Objections

Here is the first.

> You just maintained that, if we insist that *will* is scopeless, then from the outset, we eliminate "openness". But this is false. Suppose, as you grant, that "openness" is the state of affairs that obtains with respect to *p* (in n units of time) when it is not true that F*np* and not true that Fn~*p*. Well, we can maintain scopelessness and Will Excluded Middle (F*np* $\vee$ Fn~*p*) consistently with openness thus defined: we can say that neither such disjunct is true, but that the disjunction is true.

Granted. Strictly speaking, we do not *eliminate* openness thus-conceived: we simply make its expression difficult to understand. Consider, after all, the following parody of the above speech:

> You just maintained that if we insist that *necessity* is scopeless, then, from the outset, we eliminate contingency. But this is false. Suppose, as you may grant, that contingency is the state of affairs that obtains with respect to *p* when it is not true that Necessarily *p* and not true that Necessarily ~*p*. Well, we can maintain 'Necessarily *p* $\vee$ Necessarily ~*p*' consistently with contingency thus defined: we can say that neither such disjunct is true, but that the disjunction is true.

And parallel claims may be made for the other given items in (ii)–(v). But the response to any such claim is clear. We simply have no need to say that *contingency* is the state of affairs that obtains with respect to *p* when, roughly, it is indeterminate whether it is necessary that *p* or instead necessary that ~*p*. We have the theoretically far more satisfactory option of saying that it is the state of affairs that obtains when both such claims are *false*—and thus when the given disjunction is false. And so similarly for openness. Philosophers attracted to the "open future" have felt the need to invoke the mysterious sort of "openness" at issue in the first speech precisely because they have felt the need to respect (something like) scopelessness. Once scopelessness is denied, however, then we are in position to say that openness is no more mysterious than contingency, uncertainty, mere permission, agnosticism, and indecision—intuitively, all of which obtain when both of the relevant claims are *false*. Once again, we have continuity where other approaches must see discontinuity.

The second objection is related to the first:

> Well, those philosophers were on to something. For your core idea, developed at length above, is that we are inclined to accept scopelessness—and Will Excluded Middle—only because we bring with us a certain *model* of the future, viz., a model on which there is an "actual future history", the existence of which makes the distinction between ~Fn*p* and Fn~*p* practically irrelevant. But this is false. For *even if* there is no actual future history, Will Excluded Middle *still* seems true. That is, even if I am explicitly taking into account that there is no "actual future", 'Trump will be impeached in an hour or Trump will not be impeached in an hour' *still* seems true—even if, as you say, Trump is impeached in an hour on some branches, and not on others, and there is nothing at all to break the tie. And so your claim that the purported scope distinctions are being *masked* by this assumption is false: for the intuition that there are no such distinctions survives the explicit denial of that assumption.

My response to such an objection is simple: No it doesn't. It *does not* still seem that 'Trump will be impeached in an hour or Trump will not be impeached in an hour' is true, once we have before us a model on which Trump is impeached in an hour on some branches and not on others, and with nothing at all to break the tie. For consider the claim that Trump will be impeached in an hour. We check: there is nothing in the model to make such a claim true; and if a claim isn't true, it is false. So that claim is false. And consider the claim that Trump will not be impeached in an hour. We check: there is nothing in the model to make such a claim true; and if a claim isn't true, it is false. So that claim is false. But the disjunction of two falsehoods is false. Surprise! 'Trump will be impeached in an hour or Trump will not be impeached in an hour' is false. (I did promise an error-theory.) So what the objector says still seems true does not still seem true.

But the objector may wish to interject:

> But you are simply *assuming* bivalence. If we *assume* bivalence, then the given disjuncts are going to turn out false, and the disjunction false. But you cannot simply *assume* bivalence in this context.

Such an objection shows—or would show—that we have lost sight of the long-standing historical discussion of the problem of future contingents. Traditionally, the problem for the open futurist has not been that she has simply *assumed* bivalence—indeed, the problem has been that she *cannot* assume bivalence. The problem has been, in other words, that if we assume bivalence, the open futurist's position ends in contradiction (or some other similar disaster).[21] It cannot be a *problem* for my view that I am assuming bivalence: what must be shown is that something absurd follows from such an assumption, together with the denial of the claim that either such disjunct is true. And this is what I claim has not been shown.

However, the objector may wish to say more:

> Let me back up. The problem is how you are *proceeding* when evaluating 'Trump will be impeached in an hour or Trump will not be impeached in an hour'. You are simply going to the model, checking the first disjunct against that model, then returning to the model, and checking the second disjunct against that model—and you are then employing the standard semantics for disjunction to return the claim that the given disjunction is false. But you are hereby missing the intuition, which is that *looking directly at the disjunction as a whole*, the disjunction seems true—and, again, *still* seems true, even though we recognize that there is nothing in the model to support either disjunct.

But the problem here is twofold. First, I seem to stand accused of employing the otherwise perfectly standard way of evaluating the truth of a disjunction to evaluate the truth of this particular disjunction. This is not, I believe, a compelling objection. Second, the problem once more comes in the final line: why should we *still* maintain that the disjunction seems true, once we recognize that the model supports neither disjunct? Why not instead conclude that a claim that initially seemed true is not in fact true, given that model? The open future has surprises; this much, however, is not surprising.

What this sort of objector likely has in mind, however, is something like this.

> But look: *no matter how things go*, Trump gets impeached in an hour, or does not get impeached in an hour; accordingly, he will get impeached in an hour, or he won't get impeached in an hour. As we might say: it will be one or the other! How can we *deny* that it will be one or the other, when those are *the only two options?* Accordingly, even when we explicitly recognize that there is nothing to break the tie, and so it is not true that it will be one, and not true that it will be the other, we *still* must grant that it will be one or the other.

And it is here that we encounter, perhaps, the crux of the issue—and it is here, I contend, that we must be extremely careful. For how should we interpret the

[21] In particular, the traditional problem is that, if we say that both such disjuncts are false, we will have to say that the given disjunction is false—but the disjunction is an instance of LEM, and so we must deny LEM. Response: as I argued above, the disjunction is not an instance of LEM!

crucial claim here, viz., that it will be one or the other? How should that intuition be made more precise?

Prima facie, it seems that we should write 'It will be one or the other' as follows: 'Fn($p \vee \sim p$)'. And here we encounter the beginnings of what I believe to be a plausible error-theory for why 'Will Excluded Middle' '(Fn$p \vee$ Fn$\sim p$)' can *seem* so plausible, even if we grant the openness of the future. (I defend this theory at greater length in Chapter 4.) And that is that we are mistaking a true claim, viz. 'Fn($p \vee \sim p$)' with a false claim, viz. '(Fn$p \vee$ Fn$\sim p$)'. The former is not a future contingent, but a future necessity (($p \vee \sim p$) holds on *all* branches). The latter, however, is the disjunction of two future contingents. The former says, in short, that it will continue to be in n units of time that LEM holds—and surely it must, and so surely it will. The latter, however, says not that LEM will hold in n units of time, but that, one the one hand, it will be that p in n units of time, or, on the other, it will be that $\sim p$ in n units of time. And this is to say something much stronger than the former claim. On reflection, however, I believe that, often, when we try to *justify* 'Fn$p \vee$ Fn$\sim p$', we lapse into what is in fact not a justification of *that* claim, but instead a justification of 'Fn($p \vee \sim p$)'. Consider, after all, the objector's final line: even though it is not true that it will be one, and not true that will be the other, we *still* must grant that it will be one or the other. And surely that is right: it will be one or the other. No matter which future we choose, that future has it (in n units of time) that p or has it that $\sim p$—accordingly, it will be in n units of time that $p \vee \sim p$. That, I believe, is the intuition that must be respected. But that is an intuition my account can happily accommodate. And once again, a comparison with the operators at issue in (i)–(v) is instructive. In these cases, 'M($p \vee \sim p$)' does not imply 'M$p \vee$ M$\sim p$'. Similarly, I claim, 'Fn($p \vee \sim p$)' does not imply '(Fn$p \vee$ Fn$\sim p$)'. Once again, my account sees continuity where others must see discontinuity.

3.10 Interim Conclusion

It is worth summing up the picture that results from the above discussion. In sum, when we have p in n units of time on some but not all branches, and no 'actual future', we have the following:

Fn$p \vee \sim$Fnp – true.

This is a classical instance of LEM, and the second disjunct is true. Now consider:

Fn$p \vee$ Fn$\sim p$ – false.

This is not an instance of LEM, and both disjuncts are false. Will Excluded Middle is denied.

$\mathrm{Fn}(p \vee {\sim}p)$ – true.

Even if p in n units of time isn't on all branches, $p \vee {\sim}p$ certainly is, and even if there is no 'actual branch', and so no unique actual branch on which q, if q is nevertheless on *all* branches in n units of time, this should suffice for the truth of F*nq*.

This picture is plainly *simple* and it is plainly *classical*. I do not hereby claim that this is a decisive advantage for this view—but I do contend that it is a view that deserves serious consideration by philosophers working on these topics. The primary obstacle to this view has been the suspicion that its crucial resource—the distinction in scope between ~Fn*p* and Fn~*p*—is, in MacFarlane's words, simply "missing". Above, however, I have argued that this distinction is not missing, but is simply being *masked* by our implicit assumptions about the future—an argument that gains substantial traction once we see *will* as continuous with other so-called "neg-raisers". Such a view can allow us to see to *will* as semantically continuous with the modals at issue in (i)–(v) above—and thus as having meaningful scope interactions with negation—and can see *openness* on analogy with (*inter alia*) contingency and mere permission. To be sure, whether these scope interactions are *important* or *practically relevant* is a question beyond the scope of this chapter—for this question is, as I have argued, inevitably and finally a question for the metaphysician.

3.11 Supervaluationism: A Comparison

At this stage, I wish to make a comparison between my own view and the so-called *supervaluationist* view in the context of future contingents. Now, the supervaluationist maintains that any given instance of WEM is not only true, but, in their terminology, *supertrue*. I am thus in the position of denying what is, according to my opponents, not only *true*, but supertrue! It is thus worth bringing out how the supervaluationist and myself arrive at these different results. (My co-author and I discuss supervaluationism at greater length in Chapter 8.) Using a standard, simplified example, we can articulate the supervaluationist's reasoning in favor of WEM as follows:

> Suppose we have two total candidates for actuality, W1 and W2. According to W1, there is a sea-battle tomorrow. According to W2, there is no sea-battle tomorrow. This is other words for saying: if W1 is the actual world, then there will be a sea-battle tomorrow. And if W2 is the actual world, then there will be no sea-battle tomorrow. But W1 and W2 are our only candidates for actuality. Accordingly, *one* of them is the actual world. But since if W1 is the actual world, there will be a sea-battle tomorrow, then if W1 is the actual world, there

> will be a sea-battle tomorrow or there will be no sea-battle tomorrow. And since if W2 is the actual world, there will be no sea-battle tomorrow, then if W2 is the actual world, there will be a sea-battle tomorrow or there will be no sea-battle tomorrow. Thus, regardless of whether W1 is the actual world or instead W2, there will be a sea-battle tomorrow or there will be no-sea battle tomorrow. In that sense, 'There will be a sea-battle tomorrow or there will be no sea-battle tomorrow' is *supertrue*—assuming that a given world is actual, that claim is true *no matter which* world is actual.

The crucial posit of the supervaluationist is thus that *there is an actual world.* Without the assumption of there *being* an "actual world", the supervaluationist's reasoning in favor of the supertruth of WEM cannot so much as get off the ground. For I can of course grant the following: *if* W1 is the actual world, then WEM holds. And *if* W2 is the actual world, *then* WEM *also* holds. So? On my account, *neither* are the "actual world", because, again, there just *is no* "actual world" in the first place. For the supervaluationist, however, it is simply *indeterminate* (in this scenario) which world is actual—W1, or instead W2. But there is an actual world. For me, however, there just *is no* actual world.

Is this some kind of absurd result? It isn't. Suppose we have only two entities in existence, a red ball and an orange ball. Well, then the red ball and the orange ball are the only two candidates for *being President of the United States*—but we should hardly conclude that therefore one of them *is* President of the United States. To be sure, they are the only two candidates for the role—but the role, in that scenario, is simply *empty.* Similarly, I grant that W1 and W2 may be the only candidates for actuality. But I do not thereby conclude that one of them *is* actual—for, as I see it, the role of *being the actual world* is currently *empty.* (Granted, from the standpoint of the end of time, it will be filled, but that is not to say that it *is* filled *now.*[22])

In this light, we must attempt to get clear on which model, as described in Chapter 2, the supervaluationist endorses. Consider the following quote from Sven Rosenkranz: "The Ockhamist allows... while both the Peircean and the Supervaluationist Indeterminist deny... that there is a thin red line marking out the one and only course of events, of all the possible future ones, that is going to unfold" (2012: 625–626). But if the above fairly represents the supervaluationist's reasoning, then the supervaluationist *does* believe that there is a 'thin red line' marking out the one and only course of events that is going to unfold. (And if it

[22] Cf. this apt comment from Dale Tuggy, making the same realization in a slightly different context: "A couple of interesting things follow from this picture. First, there is at present no actual world!... one can reason about possible and impossible worlds, which would be maximal branches through the tree, but there won't now be any actual world" (Tuggy 2007: 33). See also Kodaj 2013 for an extended development of this point. Note: on my picture, it is not quite right to say that there is only an "actual world" from the standpoint of the end of time; there may *come* to be an "actual world" if all the indeterminism in the world is eliminated. At *that* point, God could, *ex hypothesi*, simply *deduce* which world is "actual" (Todd 2016a: 786).

doesn't, then I'm afraid I don't understand the supervaluationist's reasoning.) What they *add* is that it is *indeterminate* which course of events—which *world*—is marked out in this way. The point here is simply that the supervaluationist adopts a model on which there is a unique actual future. And as we have seen above (and in Chapter 2), any such model vindicates WEM.

Here is a comparison (explored also in the next chapter). Consider Stalnaker's supervaluationism in the context of counterfactuals—in particular, in the context of his way of preserving "Counterfactual Excluded Middle". Roughly, both Stalnaker and Lewis agree that whether a counterfactual is true is a matter of whether the closest worlds at which the antecedent is true are worlds at which the consequent is true. But in certain cases, it may reasonably be supposed that certain differing worlds are *tied* for "closeness" in the relevant way. It is here, however, that Stalnaker assumes that there is indeed a unique closest world—but simply contends that, in the relevant cases, it is *indeterminate* which world is closest. The parallel here in the case of future contingents is obvious. The supervaluationist assumes that there is indeed a unique actual future—a complete "actual world"—but simply contends that, in the relevant cases, it is *indeterminate* which world is actual.

Having said all of this, why not go supervaluationist? Well, my primary motivation is again ontological or metaphysical: *the actual world* must earn its keep—but it doesn't. It just isn't *needed*. There just is no *need* to say that there is a unique actual future, but it is indeterminate which it is. *The actual world* is an ontological or metaphysical posit that can be dispensed with. Second, the supervaluationist must deny bivalence, and must maintain that a disjunction can be true while neither disjunct is true—and both results are associated with various costs. Of course, on these latter points, we might appeal to a different context in which the supervaluationist method has been deployed: vagueness. Can we understand the claim that 'Jones is bald' is not true, 'Jones is not bald' is not true, and yet 'Jones is bald or Jones is not bald' *is* true? If we can, then why can't we understand the claim that neither Fn*p* nor Fn~*p* is true, and yet 'Fn*p* ∨ Fn~*p*' nevertheless is true?

This is a good challenge to which I do not have a fully satisfactory answer—and this is, at least in part, because I do not have anything approaching a satisfactory answer to the problem of vagueness. More generally, perhaps the phenomenon of vagueness does indeed force us to deny bivalence; this is, of course, an enormously contentious and difficult issue to which I cannot, in this book, even do minimal justice. My claim is simply that, even if we have reason from the phenomenon of *vagueness* to deny bivalence, we do *not* have such reason from considerations arising from the openness of the future. That we have such reason would follow from the one domain to the other only if the "indeterminacy" involved in vagueness were the same sort of "indeterminacy" involved in future contingents. But this is implausible. *Prima facie*, the openness of the future is not any kind of indeterminacy as that involved in vagueness. It arises out of nothing like semantic indecision, or the phenomenon of borderline cases. The openness of the future does not arise, say, because it is indeterminate whether the event that will happen tomorrow counts as a genuine sea-battle—rather than, say, a sea-skirmish. Perhaps it is a vague matter how many (or what size) ships are needed to constitute a "sea-battle". That may be so—that does seem to be so—but one thing that *is* clear is that this has nothing whatever to do with the traditional problem of the open future.

More particularly, the sorts of "indeterminacy" at stake in these domains seem fundamentally different. Indeed, if someone suggests that it is *indeterminate* whether there will be a sea-battle tomorrow, there is indeed a sense I can attach to this expression: it is indeterminate whether the event that shall be taking place tomorrow does or does not count as a "sea battle". (Or, perhaps, it is indeterminate whether the event in question will be taking place "tomorrow"—perhaps because the event partially straddles tomorrow and the day after.) However, if we simply stipulate precisely what we mean by a 'sea-battle' and 'tomorrow', the problem of the open future still arises, viz., the problem of what we should say about 'There will be a sea-battle tomorrow', when *some* ways things could unfold include precisely that kind of event during the relevant span of time, and some ways do not, and "nothing to break the tie". If *this* is the problem under discussion, however, then I do not understand (or cannot easily understand) the sense in which it may be said to be *indeterminate* whether there will be a sea-battle tomorrow. I should instead be inclined to say that it is *undetermined* whether there will be—or not yet *settled* whether there will be, or something such as this. However, the way in which an event may be "undetermined" in this sense has little or nothing to do with the phenomenon of indeterminacy as it arises in the literature on vagueness.

3.12 The Past and the Future: A Comparison

Recall the central objection to *presentist* versions of the open future I tried to address in Chapter 1. On this objection, if the future is "open", then so is the past—which it isn't. Now, a comparison between *was* and *will* has been used before to object to the thesis that future contingents are systematically false. In criticizing Charles Hartshorne's (1965) defense of this position, Stephen Cahn writes:

> Now, Hartshorne asserts that it is false that a sea-fight will take place tomorrow and false that a sea-fight will not take place tomorrow. If we represent the proposition "a sea fight will take place tomorrow" by p, then Hartshorne seems to be affirming that p is false and $\sim p$ is false. But this is surely to deny the law of

> contradiction, for *p* and ~*p* are certainly contradictories, and as such, one must be true and the other false.
>
> Here Hartshorne affirms that, in the case of a statement affirming or denying the occurrence in the future of a contingent event, it and its denial are not contradictories, but contraries such that both may be false, though not both may be true. At this point it is no longer clear what Hartshorne means by a contradictory or what he means by affirming that one proposition is the contradictory of some other proposition. Since the propositions "there was a sea-fight" and "there was not a sea-fight" are contradictories, so are the propositions "there will be a sea-fight" and "there will not be a sea-fight". (1967: 63)

Cahn's reasoning in this passage is seductive. It is also, I contend, mistaken.

The reason Cahn's reasoning is *seductive* is the following. Nearly all of us—myself included—agree that, for the propositions "there was a sea-fight yesterday" and "there was not a sea-fight yesterday", "one must be true and the other false". And this can distract us from the central issues at stake. For consider. Suppose we *denied* that "there was a sea-fight yesterday" and "there was not a sea-fight yesterday" are contradictories. If we made such a denial, I expect Cahn would be prepared to say: "So you *deny* that, of those two propositions, one must be true, and the other false? But that is absurd." In other words, if we deny that those two propositions are contradictories, this seems to commit us to the possibility—in some sense of possibility—that both are false. But we are not inclined to think that both such claims could be false. Thus, if Cahn made the given rejoinder, Cahn would thereby have a point, or a kind of a point. For it may indeed be absurd to deny that, of those two propositions, one must be true and the other false. But what is crucial here is what Cahn's discussion ignores: the *grounds* of this absurdity—or, in other words, the *sense in which* it "must" be that one such proposition is true and the other false. For I agree that there are such grounds. However, those grounds are *metaphysical*, not semantic. Cahn, however, needs those grounds to be *semantic*. But they are not.

Look at it this way. Cahn needs it that "there was not a sea-fight" is the *semantic* contradictory of "there was a sea-fight". That is, the needed claim is that it is in virtue of semantic competence that we can see that if we treat "there was a sea-fight" as *p*, then we can treat "there was not a sea-fight" as ~*p*. Cahn perhaps expects us to assent to this claim, I take it, precisely because he expects that no one will insist that there is an important semantic distinction between ~**P**n*p* and **P**n~*p*. But there is indeed such an important distinction—or so I wish to contend. It is not a common distinction to make, but that is because the doctrine that makes it salient—the open past—is not a common doctrine. On my view, however, it is not in virtue of pure semantic competence that we can "move" from ~**P**n*p* to **P**n~*p*. It is, rather, in virtue of semantic-cum-*metaphysical* competence—that is, competence with the prevailing (and, I think, clearly true) metaphysical *theory* of the past, namely, that we have a privileged past! In other words, when I consider

> **P**n*SF* ∨ **P**n~*SF* ("there was a sea-fight yesterday or there was not a sea-fight yesterday")

I do in fact accept this claim. But I accept this claim on *metaphysical grounds.* I accept this claim *because* I accept that *there is a unique actual past.* (Indeed, when it comes to the past, I accept the past-directed analogue of model (I). Model (I) is great for the past, but bad for the future.) Thus, I accept this claim *not* because, as Cahn would have it, it is an instance of LEM, or is some trivial application of LEM. This disjunction *is no instance of* LEM. The reason for accepting (what we might call) Was Excluded Middle is neither semantic nor logical. The reason is metaphysical. We have a complete, privileged past. And if we have a complete privileged past, then either, along that privileged past branch, n units of time ago, you have it that *p*, or, along that privileged branch, you have it that ~*p*. That is what it is for the given "past" to be *complete.* If we have a privileged past, then clearly one of the two given disjuncts is going to be true (although perhaps we don't know which) and the other false. If the past were open—if there were no given past history that is uniquely our own[23]—however, then Was Excluded Middle would be false (in the relevant instance)—although Excluded Middle itself certainly would *not* be false. But this is just to say that if "there was a sea-fight yesterday" is represented as *p*, we cannot simply semantically represent "there was not a sea-fight yesterday" as ~*p*.[24] This is precisely to ignore the distinction between ~**P**n*p* and **P**n~*p*.

Consequently, Cahn's argument against Hartshorne's position fails. Cahn expects us to agree that the two past tense propositions are contradictories; he then expects us to agree that, if those propositions are contradictories, then the two future tense propositions are contradictories. Cahn is right in this much: they both are, or they both aren't. My answer to this challenge is simply to say that, in fact, the pair of past tense propositions are *also* not contradictories. At the very least, we must be very careful when treating the given past tense propositions as "contradictories"—at least, we must be careful to say what *kind* of "contradictories" these propositions are taken to be. If we simply say two propositions are "contradictories" just in case one must be true and the other false, then we must be clear on the modal force of this *must.* Again, I agree: when we have **P**n*p* and **P**n~*p*, one must be true and the other false. But this is *must* is metaphysical: it is the same

[23] For a recent defense of this startling view, see Dawson 2020.

[24] Well, one cannot do so unproblematically; as I maintained in Chapter 2 with respect to *will*, one could make a parallel case that "there was not a sea-fight yesterday" is *ambiguous* between ~**P**n*SF* and **P**n~*SF*—the default reading of course being the latter. I set this issue aside.

must at issue in saying that there *must* be a privileged past. However, because these claims are (as it were) *metaphysical* contradictories, it does not follow that they are *semantic* contradictories. And if these claims are not semantic contradictories, neither are their future tense counterparts. At any rate, if these pairs cannot both be false, then this is one's *metaphysics* talking—not one's pure semantic competence.[25]

3.13 No Fact of the Matter?

By way of concluding this chapter, let me try to head off the following reasonable misgivings about the picture that results from the above discussion. Arguably, when we are attracted to the open future, the intuition to which we are attracted is that, given the relevant openness, there is something about which there is *no fact of the matter*. But precisely this language—that there is no fact of the matter!—strongly suggests an intuition that must be given a non-classical interpretation, if it is given one at all. For if there is no fact of the matter as concerns p, this strongly seems to imply that it is neither true *nor false* that p—after all, if it were simply *false* that p, there *would* be a fact of the matter as concerns p. Namely, it is false! Accordingly, by insisting that future contingents are simply *false*, we have arguably abandoned precisely the core set of intuitions that attracted us to the open future to begin with.

The misgiving is misplaced. As is the case for so much of our language in this area, the language of there being "no fact of the matter" is difficult language. Plausibly, however, we do not need to abandon bivalence to appropriately speak of contexts in which there is no "fact of the matter".[26] Consider incomplete fictions. There is no fact of the matter whether Gandalf put on his left shoe first or instead his right the day he first met Frodo; this means that it is false that, in the fiction, he put on this left shoe first, and false that, in the fiction, he put on his right shoe first.

[25] My claim here is that, if one regarded the past as "open" (i.e., that "past contingents" aren't true), it would be reasonable to treat "past-contingents" as simply *false*. In other words, I wish to treat *was* and *will* as *semantically* on a par—the difference is solely metaphysical (there is privileged past, but no privileged future.) There is, however, one serious complication with this argument: this contention would seem to commit me to the view that *was* is similarly a modal, viz., a universal quantifier over past branches. And whereas linguists certainly do sometimes treat *will* as a modal, to my knowledge, no one has ever treated *was* as a modal. I am not entirely sure what to say about this issue, but my current feeling is this. It is indeed plausible to say that there is (and always has been) a covert modal component to *was*. To say that it was the case that p is indeed to say that, in all of the *available pasts*, p. However, as in the above, we can explain why no one tends to sense a difference between ~Pnp and Pn~p by appeal to the obvious fact that we systematically tend to assume that there is only one available past—in which case, the distinction between ~Pnp and Pn~p is practically irrelevant.

[26] For one treatment of this issue, see Azzouni and Bueno 2008. As Azzouni and Bueno note, Quine famously maintained that there is no fact of the matter about whether, in Junglese, "gavagai" means "rabbit" or instead "undetached rabbit parts"—and yet Quine was still a strong proponent of bivalence. As they note, the development of this position does require substantial care. Personally, I am tempted toward the view that there is no fact of the matter about the "correct" usage of "no fact of the matter".

Nothing needs to be neither true nor false in this scenario. Or consider cases of *moral ties*. Three charities are all equally good. Jack must donate to one of them. However, there is no fact of the matter concerning to which charity he must donate. This means that it is not the case that he must donate to charity 1, not the case that he must donate to charity 2, and not the case that he must donate to charity 3—although, of course, he must donate to charity 1, or charity 2, or charity 3. Again: nothing needs to be neither true nor false in this scenario. And yet it seems like a scenario in which there is no fact of the matter concerning to which charity Jones must donate.

Similarly: there is no unique actual future; as a consequence, there is no fact of the matter concerning whether the event *will* or instead *will not* occur. Nothing needs to be neither true nor false in former scenarios, and nothing needs to be so in the latter as well. Consider, after all, how we might put the intuition in question: there is no fact of the matter whether the events will happen. If we say that future contingents are false, have we taken back what we said? We have not. And how could we have? We certainly haven't said that the given events *will* happen. Nor have we said that those events *will not* happen. Precisely the indeterminacy we originally postulated is retained. And yet bivalence is retained as well. The result is a neglected picture of a neglected doctrine. The result is the open future, classical style.

4
The Will/Would Connection

In this chapter, I am going to express some opinions about the relationship between the theory of *will* developed above with various theories of the counterfactual conditional, 'If it had been the case that p, it would have been the case that q'. I am going to try to do this without turning this book into a book on, or about, the counterfactual conditional. My project is simply to display that my own position concerning future contingents is deeply similar to a more familiar position concerning counterfactuals—namely, a position that treats a relevant subclass of counterfactuals as *false*, and accordingly denies what has been called "Conditional Excluded Middle":

> (CEM) If it had been the case that p, it would have been the case that $q \vee$ If it had been the case that p, it would have been the case that $\sim q$.

CEM has been a matter of longstanding controversy in semantics and metaphysics. My goal in this chapter is to bring out the ways in which denying CEM is parallel to denying Will Excluded Middle. I am (perhaps painfully) aware of how much a minority position it is that I have defended in Chapters 2 and 3. There is something nice about novelty—but not too much novelty. Thankfully, once we see how others have denied CEM, we can see how my treatment of future contingents is not as strange and unexpected as it initially seems. Moreover, given the history of the debates over CEM, I can, in this chapter, let others do much of the talking. And that is what I plan to do.

4.1 Grounding

Recall the claim of Chapter 1: future-tense truths must be grounded in present conditions and laws. The first and perhaps most obvious point of contact between future contingents and counterfactuals is that many have argued that counterfactuals must be grounded in some analogous way—not necessarily in the present, but in (something like) the *actual*, or what has been called the "categorical". Hence Sider:

The Open Future: Why Future Contingents are All False. Patrick Todd, Oxford University Press. © Patrick Todd 2021.
DOI: 10.1093/oso/9780192897916.003.0005

> Second example: brute counterfactuals.[1] Most would say that when a counterfactual conditional is true, for example 'this match would light if struck', its truth must be grounded in the actual, occurrent properties of the match and its surroundings. Someone who postulates counterfactuals not grounded in this way is Alvin Plantinga (1974: 180). Imagine God deliberating whether to create a certain free creature, C. According to Plantinga this amounts to deciding whether to cause a certain individual essence to be instantiated; the essence exists whether or not instantiated. God must take into account certain true counterfactual conditionals specifying what free choices C would make if placed in certain circumstances. These counterfactuals hold even if God decides not to create C, and therefore seem objectionably ungrounded, since they depend in no way on what existing things are like. (2001: 40)

Plantinga certainly denies that counterfactuals must be "grounded" in the sense at stake.[2] However, I take it that the grounding claim applies more widely than merely to this critique of Plantinga; according to the grounding claim (as I am conceiving it), even some counterfactuals about *actual* objects may lack proper "grounds". Here we shall take as our canonical example the flipping of a perfectly fair, indeterministic coin. What if I had flipped such a coin on my 35th birthday? Would it have landed heads, or instead tails? (Here I treat any non-heads as a tails.) According to the grounding claim, neither of the following counterfactuals have proper grounds, and therefore both fail to be true:

> If I had flipped this fair, indeterministic coin on my 35th birthday, it would have landed heads.
>
> If I had flipped this fair, indeterministic coin on my 35th birthday, it would have landed tails.

This is surely not *all* the grounding claim implies (indeed, one worries that, pressed too hard, it is going to imply too much), but if it implies anything, it

[1] This is Sider's second example of metaphysical views that (in his words) *cheat*—that is, help themselves to truths for which their ontology provides no "truthmaker" (or perhaps nothing on which to "supervene"). As in Chapter 1, I am skeptical that the best way to object to a given view of counterfactuals is via a *general* claim about truth, but I set this point aside.

[2] Hence Plantinga's (1985: 378) famous retort to the so-called "grounding objection" to Molinism: "It seems to me much clearer that some counterfactuals of freedom are at least possibly true than that the truth of propositions must, in general, be grounded in this way." Others defending Molinism agree that such counterfactuals cannot be grounded, but take this as a reason to abandon the grounding requirement; see, e.g., Otte 1987, Freddoso 1988: 68–75, Craig 2001, and Merricks 2007: 155. (For a treatment of whether the grounding requirement applies equally to claims about the future as it does for counterfactuals, see Hunt 1990.) For a recent (and particularly stark) rejection of the grounding claim *not* motivated by Molinism, see Stefánsson 2018. See also Hawthorne 2005b: 405 for a similar proposal.

implies that neither of the counterfactuals at issue here are true. That is good enough for present purposes.

But now the point. The disjunction of these claims is an instance of CEM. According to the grounding claim, neither of these counterfactuals is true. According to bivalence, they are therefore false. If they are false, then CEM is false.[3] We thus can take our pick between any two of the following three theses, but together they are mutually incompatible:

The Grounding Claim
Bivalence
CEM

Plantinga maintains bivalence and CEM, and denies the grounding claim. Stalnaker accepts the grounding claim and CEM, and denies bivalence; on his view, neither of the above disjuncts is true, but the disjunction is true. Finally, Lewis and Williamson accept the grounding claim and bivalence, and deny CEM. Since they do not also wish to deny the Law of Excluded Middle, such theorists defend a scope distinction between $\sim(p > q)$ and $(p > \sim q)$. (I use '>' throughout to indicate the counterfactual conditional.) Such are our options.

Thus far, it is obvious how similar this dialectic is to the problem of future contingents described above. In that case, there is a grounding claim, a result that neither one of Fnp nor Fn$\sim p$ is true, and a subsequent conflict with bivalence and Will Excluded Middle (WEM). In both cases, I accept the grounding claims, and do not see a compelling reason to deny bivalence to preserve WEM and CEM, and accordingly deny both.[4] It would certainly help my case, however, if some of the strategies we may employ to explain away CEM also helped us to explain away WEM. And, on inspection, that is just what we find. To Williamson's discussion of CEM I now turn.

4.2 Williamson on Conditional Excluded Middle

Here is Timothy Williamson, a chief defender of bivalence, and (not incidentally) a chief critic of Conditional Excluded Middle:

> The only question, I think, is whether there are good reasons for independently accepting conditional excluded middle (CEM). One could accept (CEM) even if

[3] This is, of course, assuming that no one will want to say that the disjunction could be true although both disjuncts are *false*. I do not address this possibility, interesting though it is.

[4] As I said in the Introduction, my point here is not so much that bivalence is *true*, but that preserving WEM and CEM (and the grounding claim) is not a compelling reason to deny it. WEM and CEM can go.

> one thinks that the negation of a counterfactual is different from the result of negating its consequent. (CEM) is a principle that has been defended by Stalnaker. It is valid on his logic of counterfactuals, and he has defended it without any confusion about what the contradictory of a counterfactual conditional is. In my view, his defence is unconvincing and unnecessary. We have perfectly good logics for counterfactuals, like the one David Lewis gave, where (CEM) is invalid. The cases to consider are cases where the antecedent seems to be in some way completely neutral between the consequent and the negation of the consequent, maybe cases where, say, indeterminism holds. If we have 'If the coin had been tossed, it would have come up heads' and 'If the coin had been tossed, it would have could come up tails' and therefore 'not heads', there is nothing to choose between them. It's not that we don't know which is true, but as it were, reality itself doesn't decide in favour of either of them. One way we might think about this is that in order for a counterfactual conditional to be true, there has to be some sort of connection between the antecedent and the consequent. Let's not now try to specify what sort of connection that is required. It seems that there would be cases where the antecedent lacks that connection to the consequent, but also lacks that connection to the negation of the consequent. In those cases conditional excluded middle would fail, but that is not a failure of bivalence, because if what's required for a counterfactual conditional to be true is that there is a connection of the right kind between the antecedent and the consequent, then all that is required for it to be false is the absence of such a connection. It is not required that there be some alternative connection going the opposite way. We could compare them to existential claims. What's required for 'There is a talking donkey' to be true is just that there be such a donkey. That means that what is required for that sentence to be false is simply that there be no such donkeys. It is not required that there be some other kind of donkey, that prevents all other donkeys from talking. It is only required that there be no talking donkeys at all. Similarly, if a counterfactual informs us that there is a connection of a certain kind, what it amounts to for it to be false is simply that there be no such connection.
>
> (Williamson and Antonsen 2010: 22–23)

There is much to appreciate in this fine passage.

First, note the procedure Williamson employs. We consider a case in which the situation described in the antecedent is *neutral* with respect to the consequent and its negation. The intuition is then that it would *metaphysically arbitrary* if reality somehow decided on behalf of the antecedent that, well, *this* consequent rather than *that* would be what obtained if the antecedent had obtained. What is not fixed by the scenario described in the antecedent is simply not fixed at *all*. The intuition here is very similar to the intuition that where what exists in the present is *neutral* with respect to an event's happening or failing to happen, it would be metaphysically arbitrary if reality somehow decided on its behalf to supply a further fact—the fact that, well, *this* event is the one that is going to obtain.

Second, note Williamson's contention that what is required for the falsity of a counterfactual is simply the *absence* of a certain connection, not the *presence* of an alternative connection going the other way. Can we make a parallel point in the case of future contingents? Arguably we can. Say that when something is such that it is *going to happen*, it thereby has a certain positive status. What is thereby required for the falsity of a future contingent is simply that the event it describes should *not* have this positive status. It is not required that the event in question have an alternative *negative* status, the status of being going to *not* happen. Of course, one might argue independently that an event's *not* having the positive status implies its having the negative status. Indeed, as I have emphasized throughout this book, one might argue that, because there is a complete "actual future history" or "actual world", an event's not having the positive status implies its having the negative one. That is not an unreasonable assumption, but my point is that it is, in any case, a *metaphysical* assumption which might coherently be rejected. The bulk of this book can be seen as an attempt to open up the *semantic* space to say that an event's *not* having the positive status does not *semantically* imply its having the *negative* status.

In earlier reflections on CEM, Williamson writes:

> Had there been nothing but a gold or silver sphere, would it have been gold? Apparently, to answer "Yes" is to say that had there been nothing but a gold or silver sphere, it would have been gold; to answer "No" is to say that had there been nothing but a gold or silver sphere, it would not have been gold (and therefore would have been silver). Not only have we no reason to say either of these things, but it is hard to take seriously the supposition that one or other of them is, unbeknownst to us, true. Can subjunctive conditionals then fail to be either true or false, thereby falsifying the principle of bivalence?
>
> There is an obvious way of dealing with this threat, as follows. To answer "Yes" to the original question is obviously unacceptable. To answer "no" is to deny that, had there been nothing but a gold or silver sphere, it would have been gold. That is to assert that it is not the case that, had there been nothing but a gold or silver sphere, it would have been gold.... However, none of this is to assert that, had there been nothing but a gold or silver sphere, it would not have been gold. For there is no reason why negating a subjunctive conditional should be equivalent to negating its consequent... We can also, of course, assert the triviality that, had there been nothing but a gold or silver sphere, it would have been either gold or not gold – for there is no reason why this conditional with a disjunctive consequent should entail the disjunction of unattractive conditionals, that either had there been nothing but a gold or silver sphere, it would have been gold or had

> there been nothing but a gold or silver sphere, it would not have been gold.... Hence we can reasonably answer "No". But if it is not the case that had there been nothing but a gold or silver sphere it would have been gold, then the statement "Had there been nothing but a gold or silver sphere, it would have been gold" is false; since it is false, it is bivalent. In other words, the apparent threat to the principle of bivalence is diagnosed as resulting from a confusion between the scopes of negation and [the] subjunctive conditional operator. (1988: 405–406)

These are, I hope it is clear, familiar themes. I wish to make one observation in particular. Consider Williamson's claim that "there is no reason why this conditional with a disjunctive consequent should entail the disjunction of unattractive conditionals." Williamson elaborates on this idea later:

> However, (V) $[(p > (q \vee r)) \rightarrow ((p > q) \vee (p > r))]$ is not a very plausible principle. For example, it takes us from the trivial premise that, had there been nothing but a gold or silver sphere, there would have been nothing but a gold or silver sphere to the implausible conclusion that either, had there been nothing but a gold or silver sphere, it would have been gold or, had there been nothing but a gold or silver sphere, it would have been silver. Stalnaker will say that the disjunction can be true even though neither disjunct is, but that reply is inadequate, for the disjunction itself is implausible. Moreover, the sense that (V) is implausible certainly does not depend on a prior commitment to bivalence, for Michael Dummett, who is no friend of bivalence, has objected to (V) – his example is that "If Fidel Castro were to meet President Carter he would either insult him or speak politely to him" does not entail "Either if Fidel Castro were to meet President Carter he would insult him or if Fidel Castro were to meet President Carter he would speak politely to him". (1988: 412)

Williamson denies the principle that $[(p > (q \vee r)) \rightarrow ((p > q) \vee (p > r))]$. Let us focus first on the case of where we let $r = {\sim}q$, i.e., $(p > (q \vee {\sim}q))$. Williamson's point here seems to be this. The denier of CEM can of course admit that there is a "triviality" in the *neighborhood* of CEM, but that triviality is not CEM itself. To assert $(p > (q \vee {\sim}q))$ is not to assert what we might call a subjunctive contingency, but instead a subjunctive necessity: $(q \vee {\sim}q)$ holds in *any* scenario, and therefore would *still* hold under the scenario described in the antecedent. (I set aside per impossible counterfactuals—those with impossible antecedents.) But, Williamson contends, it is difficult to see how a subjunctive *necessity* should, in itself, entail the *disjunction* of two subjunctive contingencies. So far, what we have is something very close to the position I developed in Chapter 3: I can of course admit that $\text{Fn}(p \vee {\sim}p)$ expresses a truth, but I deny that we can move from that to $\text{Fn}p \vee \text{Fn}{\sim}p$. There is no reason why a prediction of an attractive disjunction should entail a disjunction of unattractive predictions.

However, I think Williamson slightly undersells the utility of this compelling point in explaining the (at least *prima facie*) attractiveness of CEM. Arguably, those who have the intuition that CEM is valid have it at least in part because it strikes us as trivial, or otherwise *logically* true. The intuition is perhaps an intuition to the effect that p and $\sim p$ are the *only two options*. When we have only two options for how things would have been, how can we *deny* that, had things been that way, one of those two options would have been the actual one? If those are the only two options, surely we must grant that it would have been one or the other!

But wait. What are we saying when say that "it would have been one or the other"? Arguably, what we are saying *just is* that, had it been that p, it would have been that $q \vee \sim q$. That is, had it been that p, it would have one (q) or the other ($\sim q$). And that is surely right: to use Plantinga's example, had Curley been offered the bribe, it would have been either that he takes or that he does not take it. It is not as if, had Curley been offered the bribe, the consistency of the world would have failed too, so that what we would be seeing, had Curley been offered the bribe, is a situation in which we have *neither* that he has accepted it *nor* that he has not accepted it. No. Had Curley been offered the bribe, what would have been is that he takes the bribe or he doesn't take the bribe. If *this* is the intuition in question, then this is an intuition the denier of CEM can happily accommodate. But once we realize that we can accommodate *this* intuition, my claim is that it is difficult, on introspection, to tell whether one's intuition is really ($p > (q \vee \sim q)$), or instead CEM itself. The question is whether, when we have the intuition in question ("It would have been one or the other"), we can trivially insert another "if/would have": "If... it would have been one, or if... it would have been the other". On reflection, however, we must grant that *this* is at least less obvious than the former intuition. Similarly, I contend that, once we realize that we can retain "It will be one or the other!", it is not obvious that we must similarly retain "It will be one, or it will be the other!" Parallel points can be made in case of the open past. "But it was one or the other!" Yes. It was. But it was one, or it was the other? No—anyway, not if the past is genuinely open (which it isn't, but the point remains).

On a similar theme, Williamson writes:

> Stalnaker argues that "the normal way to contradict a counterfactual is to contradict the consequent, keeping the same antecedent." His argument turns... on the sheer appearance of contradiction in saying things like "Had there been nothing but a gold or silver sphere, it would have been either gold or not gold, but it is not the case that, had there been nothing but a gold or silver sphere, it would have been gold, nor is it the case that, had there been nothing but a gold or silver sphere, it would not have been gold." Lewis himself admits that there is an *appearance* of contradiction. (1988: 409)

Williamson, however, is not much impressed by this appearance—and neither am I. Again, there is a precisely parallel issue in the case of future contingents. I contend that we could have ~Fnp, and ~Fn~p, and yet Fn($p \vee \sim p$). There is, I admit, an *appearance* of contradiction here. But there is, I contend, no contradiction.

Williamson goes on:

> Stalnaker has certainly adduced evidence that we tend to *treat* the denial of ($p > q$) as though it were equivalent to the assertion of ($p > \sim q$), but he does not show that this is not a fallacy that we tend to commit, rather than something constitutive of the truth conditions of our thoughts and sentences. (1988: 411)

There is a sense in which I agree with the spirit of this remark. And I am somewhat inclined to make a parallel remark about future contingents:

> Various authors have certainly adduced evidence that we tend to treat the denial of Fnp as though it were equivalent to the assertion of Fn~p, but they do not show that this not a fallacy we tend to commit, rather than something constitutive of the truth conditions of Fnp.

It is important to note, however, that I do not think of the "move" from ~($p > q$) to ($p > \sim q$) as a kind of *fallacy*. For instance, I do not think Plantinga is committing a fallacy when he reasons from the non-truth of ($p > q$) to the truth of ($p > \sim q$). After all, Plantinga thinks that there exists (as it were) a book of primitive counterfactuals specifying what any indeterministically free agent would (and would not) do in any possible circumstance of choice—in other words, that the truth of a given counterfactual does not in any relevant way need to be "grounded in" what we have called the "categorical". Clearly, given such a conception of the "grounds" (or non-grounds) of counterfactuals, we can reason from the non-truth of ($p > q$) to the truth of ($p > \sim q$). For if the ungrounded fact of the matter is not that ($p > q$), the ungrounded fact of the matter is that ($p > \sim q$). Plantinga's point *just is* that, either way, there is some ungrounded fact of the matter here, one perhaps unknown to us, but one that is known to God. Thus, when Williamson earlier writes, "It's not that we don't know which is true, but as it were, reality itself doesn't decide in favour of either of them," Plantinga simply stops him right there. No. Reality *does* decide, and God knows what it has decided. (This is not, many have complained, an attractive picture of God.) In other words, for Plantinga, it is precisely in the nature of the book of counterfactuals to *specify* whether if p we would get q or whether if p we would instead get ~q. It is only if the truth of counterfactuals must in some relevant way be grounded in the categorical that we could not move from ~($p > q$) to ($p > \sim q$).[5]

[5] Again, Plantinga certainly isn't the *only* philosopher to take this position; as we'll see below Stefánsson (2018), for instance, preserves CEM via the postulation of primitive *counterfacts* as truth-makers for the relevant counterfactuals. Here we have a natural comparison with the *primitive future directed facts* of Chapter 2; more on this shortly.

But whereas Plantinga is perhaps making a certain kind of "mistake" in supposing that "reality decides", this is no kind of *fallacy*. The debate is instead metaphysical. The debate, in other words, is at least in part whether the truth of counterfactuals must be grounded in the categorical. But now the broader point. In ordinary life, we naturally tend to prescind from such inconvenient facts as that there is no fact of the matter about what would have happened in indeterministic scenarios that never occur—if it even *is* a fact that there are no such facts, which, as we have just seen, is disputed. My point, in short, is that most of us, even if we agree on some theoretical ground that there are no such facts, when we set down this book, we shall soon thereafter reenter the metaphysically lax environment of the everyday. And in that environment, whether there indeed are such facts is of little or no concern; we just proceed as if there were. Since we are proceeding in this lax environment, we presuppose in that environment that there are such facts; accordingly, in that environment, we tend to treat the denial of the claim that Jones would have passed as equivalent to the claim that he would have failed. If you want to *deny* that Jones would have passed, but *also* deny that he would have *failed*, this is going to be *because* you are reasoning from a *philosophical* principle to the effect that the truth of counterfactuals must be grounded in the categorical.

I would like to suggest it as some kind of principle that natural language tends to expand to accommodate the most permissive conception of what facts there are in the given linguistic community. After all, people who insist that there are no facts in a given domain of conversation will tend to find themselves left out of such conversations. And no one likes to be left out. Consider a scenario—an utterly commonplace scenario—in which a sports fan asks another, "God! Do you think we would have won if only Jones had made that catch?" Note the stark difference between the following replies:

> I don't know! On the one hand, we would have had the momentum, but on the other hand, there was plenty of time left on the clock, the game is still a chancy game, and it could still then have gone either way...

That's a perfectly respectable, cooperative reply. But then consider:

> Well, there's really no fact of the matter concerning whether we "would" have won. After all, given the indeterminism inherent in the game, there are approximately equiprobable scenarios under which Jones makes that catch and we go on to lose, and scenarios in which Jones makes that catch and we go on to win, and nothing to break the tie. But in such a circumstance, reality simply doesn't decide...

I have a great deal of sympathy for this metaphysician's speech; I also have a great deal of sympathy for this metaphysician's friend, who has been made to listen to it.

For our metaphysician has abruptly decided *not* to prescind from the inconvenient facts—there really *is no* fact of the matter concerning who would have won the game—and has done so in a context in which we were all having fun supposing there were such facts. That is not very nice. And that will tend to get you disinvited from future parties. Strategy: don't be difficult about whether there really are facts about what would have happened in the given scenarios. Just pretend that there is a book of ungrounded counterfactuals, and speculate about what is written in that book.[6]

My explanation—or part of my explanation—of the data that seems to support Conditional Excluded Middle is thus precisely parallel to my explanation of the data that supports Will Excluded Middle. Ordinary thought and talk presupposes that there *is* a fact of the matter concerning who would have won the game. In other words, ordinary thought and talk tends to treat our failure to know who would have won as *ignorance*, and *not* as failing to know what is not there to know. In effect, we have the same neg-raising explanation as that provided in Chapter 3. Unofficially, that explanation might be put as follows:

1. For any p and any q, the book of ungrounded counterfactuals has it that $(p > q)$ or instead $(p > \sim q)$. (Implicit, unspoken assumption)
2. $(p > q) \vee (p > \sim q)$ (CEM; Trivial application of (1))

[6] An important point that deserves more attention than I shall give it: a precisely parallel observation can be made in the case of future contingents. Suppose someone asks you who will be the next U.S. president. Now consider the stark difference between the following two replies:

I have absolutely no idea. (Perfectly respectable; perfectly cooperative)
There is no fact of the matter. (Perfectly asinine, even if *philosophically* respectable)

And this because, in ordinary life, *we do not tend to think of the future as genuinely open*. We treat our not knowing who will be the next president as *ignorance*, not as failing to know what is not "there" to be known. Notably, only the most pathological of even the open-futurist *philosophers* would ever think to say something like the latter in ordinary life. For my own part, I suppose I *do* think (I certainly do not *deny*) that there is no fact of the matter about who will be the next U.S. president, but I would never dream of responding in this way in ordinary life. It is only once we move into a philosophical context that I would be prepared to say (or, really, even be prepared to remember that I have a philosophical view according to which the technically correct answer is) the latter.

This observation has an important methodological upshot: we should be suspicious that we can easily prize apart the question of what is felicitous to say in the ordinary context, and what would be felicitous to say in the philosophical context in which it is taken for granted that there really *is* no fact of the matter about who will be the next U.S. president. Many philosophers are skeptical of the scope distinctions I have defended in this book; and part of the grounds of their skepticism is (or may be) the suspicion that if these scope distinctions were "there", we should be able to recover them easily and naturally (and without the determined effort I have [doggedly] displayed herein). After all, they might think, we *already* think of the future as "open", but we do not easily "hear" the distinctions I have been drawing. But this is simply false. As is evidenced by the undisputable asininity of the above reply, we do *not* "already" think of the future as open—in the relevant sense of "open" (cf. Hughes 2015: 213–215). My own suspicion is that, often, when a philosophical interlocutor does not hear (or denies that there exists) one of the scope distinctions I have defended, what is in fact going on is that that philosopher is not really taking on board the claim that the future is open; what I am in fact getting is thus not this philosophers' judgment that, even if the future is open, the relevant claim is infelicitous, but instead this philosopher's judgment that there is something very strange about thinking of the future as "open" in the first place—which is, however, a point I am ready to grant.

3. $\sim(p > q)$ (the proposition considered or uttered)
4. $(p > \sim q)$ (the stronger proposition conveyed)

It is only if you deny premise (1) that you would have cause to reject (2), and therefore the move from (3) to (4). Denying (1) amounts to the thesis that counterfactuals must be grounded in the categorical. Accordingly, if one can make plausible on *metaphysical grounds* that (1) should be denied, one can explain away the linguistic data which seems to indicate that we can move unproblematically from (3) to (4). That data is explanatorily dependent on our *already* accepting (1), and thus cannot itself constitute reason to *accept* (1). (I provide a more official characterization of the relevant neg-raising inference below.)

But back to future contingents. I deny that we can move from ~Fn*p* to Fn~*p*. However, I do not regard this move as any kind of *fallacy*. For that move is motivated by an implicit assumption: that there are primitive (with respect to the present) facts about undetermined aspects of the future. (Again, recall the *primitive future directed facts* of Chapter 2.) However, it is not a fallacy to suppose that there are such facts, any more than it is a fallacy to suppose that there are brutely true, ungrounded counterfactuals. Far from it: that supposition is perfectly reasonable, even if I disagree with it. In other words, if you were not convinced by the argument (such as it was) in Chapter 1, then I am certainly not prepared to charge you with the commission of a fallacy. I am instead prepared to say that we disagree on a number of metaphysical matters on which it is reasonable to disagree. However, if you insist that *I* am committing a kind of fallacy by *not* being willing to move from ~Fn*p* to Fn~*p*, then my response once again is to point to the metaphysics: if the metaphysics of the world includes facts about undetermined aspects of the future, and thus includes a 'unique actual course of history', then we can make the move—and not if not. But there is no fallacy involved in denying that there is such a history, even if there is a mistake.

4.3 A Brief Interlude on "Might" Arguments

Let me pause to note the following. My complaint against CEM is *metaphysical*, not semantic. However, the primary arguments considered in the literature against CEM seem to me to be *semantic* arguments—roughly, "might" arguments from the truth of a principle sometimes called "Duality". Duality maintains that "If it were the case that *p*, it would be the case that *q*" and "If it were the case that *p*, it might not be the case that *q*" are contradictories: if one is true, the other is false, and vice versa. Duality is sometimes motivated by observations about assertability. As Mandelkern notes in reviewing the case for Duality, the following is felt to be "quite odd":

(C) If the coin had been flipped, it would have landed heads; and if the coin had been flipped, it might have landed tails.

Duality, of course, would explain the oddness of (C): if it is true that coin *would* have landed heads, it is therefore false that it *might* have landed tails—and if it *might not* have landed tails, it is false that it *would* have landed heads.[7]

Though (C) is an odd thing to assert, my concern with CEM is not that *would* and *might not* can be seen to be (in this or some other way) contradictories on semantic grounds alone. My concern is that CEM would seem to commit its proponent to facts about what would have happened that go beyond facts that can be grounded in the categorical. These points have an important analogue in the case of future contingents. For note that we could develop parallel arguments for "Duality" in the case of *will* and *might not*. For instance, the following likewise seems "quite odd":

(D) There will be rain tomorrow; and there might not be.

(D) is likewise an odd thing to assert; however, the arguments of this book (against WEM) do not proceed from anything like the unassertability of (D). (It is the *Peircean*, if anyone, who maintains that (D) embodies a semantic contradiction; according to the Peircean, *will* just *means* something that would imply the falsity of the *might not* claim.[8]) Of course, I *have* defended the claim that (D) is always going to be false, and perhaps necessarily so: if the future is genuinely open, then if it is really true that there (objectively) *might not* be rain tomorrow, there could be nothing that grounds the claim that there *will* be rain tomorrow—in which case (D) is certainly going to be false. But this is a metaphysical argument for the (necessary) falsity of (D). It is not a semantic argument from its unassertability.[9]

[7] Mandelkern 2018. For discussion, see, e.g., DeRose (1994) and (1999) and Hawthorne (2005b).

[8] Consider, for instance, the argument of the arch-Peircean Charles Hartshorne. (NB: Hartshorne was a "Peircean" before Prior invented the term.) Hartshorne quotes Scrooge in *The Christmas Carol*: "Are these the shadows of the things that Will be or the shadows of things that May be, only?" Hartshorne: "There is a master of language [i.e., Dickens]. Will and May are nicely distinguished in ordinary speech" (Hartshorne and Viney 2001: 39).

[9] At this point, I'd like to casually insert one trenchant objection to the position defended in this book, developed by Stephan Torre (at a workshop on an initial draft of this book [Edinburgh, June 2019]). Torre's observation is simple. Suppose a mad scientist tells you that tomorrow he'll flip an indeterministic coin; if it lands heads, you'll get ice cream, if tails, he'll torture you for five hours. Intuitively, you now fear *that you will be tortured tomorrow*. But wait. You come to accept the theory propounded in this book; accordingly, you accept that it is false that you will be tortured tomorrow. And yet: there is certainly still something that you fear—the possibility of your being tortured tomorrow certainly hasn't been ruled out! The result thus appears to be that I must deny a plausible principle Torre has called FEAR:

(FEAR) If S fears that p and subsequently learns that p is false, then it is no longer appropriate for S to fear that p.

I am not entirely sure what to say in response to this problem, but here are two thoughts. First, I am inclined to say that the problem here simply reveals how deeply our bias is towards the view that there

4.4 Models and Semantics Once More

At this stage, it is crucial that I should bring out the close parallels between the issues discussed in this chapter, and the issues of Chapter 2—that is, the parallels between my semantics for *will* and the three "models" considered in that chapter, and a familiar semantics for the counterfactual conditional, and three parallel "models" we might consider in that domain as well.

The parallels are clear. Recall the semantics for *will* that I defended in Chapter 2:

> (AAF) It will be in n units of time that *p* iff in all of the available futures, in n units of time, *p*.

Will is thus treated as a universal quantifier over available futures. And recall the definition of the *available* futures. The available futures are those futures consistent with the past and the laws and the primitive future directed facts. We then have three different models of the undetermined future to consider. On model (I), even if there are multiple futures consistent with the past and the laws, there are primitive future directed facts, and it is determinate what they are; thus, only one given future—the unique actual future—is *available*. On model (II), the primitive

is a unique actual world. If there is a unique actual world, then when we learn that it is false that we will be tortured tomorrow, we learn that we won't be tortured tomorrow—and so there is nothing left to fear. But if there just is no such thing as the unique actual world, then even if we learn that it is false that we will be tortured tomorrow, that doesn't tell us that we won't be tortured tomorrow—we certainly could end up being tortured tomorrow, and so there is indeed something left to fear. I suppose this amounts to me simply biting Torre's bullet. At any rate, note that I can at least accept a *nearby* principle that arguably captures the relevant intuition:

> (FEAR2) If S fears that it will be that *p* and subsequently learns that it won't be that *p*, then it is no longer appropriate for S to fear that it will be that *p*.

So we're simply back to the same old scope distinction.

Second, I am inclined, once more, to make a comparison with theories of the counterfactual that deny CEM, in the hopes that my position looks no worse. And so consider the following. Suppose someone has been going around offering a bribe, but the bribe was never offered to Jack. Now, Jack's character is consistent both with his taking the bribe and his rejecting it; accordingly, had he been offered the bribe, whether he takes it would have been resolved solely by Jack's indeterministic freedom. Now, Jack comes to accept the theory of counterfactuals defended (*inter alia*) by Lewis. And suppose he says:

> Jack: I know that there's no fact of the matter whether I would have taken the bribe. But boy, I'm glad I wasn't offered it.
> B: Why is that?
> Jack: Because I fear I would have taken it.

More to the point: being a good anti-Molinist, Jack accepts that it is false that if he had been offered the bribe, he would have taken it. And yet: he is glad that he wasn't offered the bribe. And why is that? Because he fears that he would have taken it. He knows that it is false that he would have taken the bribe—and yet, he fears that he would have taken the bribe. I am not sure whether this makes sense; I suppose all I want to say—echoing the general theme of this chapter—is that my own position, once more, is no worse off than a Lewis-style position on counterfactuals. (Thanks to Stephan Torre for this objection, and to Brian Rabern for helpful discussion.)

future directed facts are indeterminate; thus, though only one future—the unique actual future—is consistent with those facts, and only one future is *available*, it is indeterminate which future this is. On model (III), there are no primitive future directed facts, and so the available futures just are the causally possible futures.

Now, (AAF) is plainly parallel to a familiar semantics for counterfactuals implicitly encountered above. On this semantics for counterfactuals, the counterfactual conditional ($p > q$) is a *universal quantifier* of some kind—viz., a universal quantifier over what we might call the "counterfactually available" p worlds (which maintains that they are q worlds).

> (ACW) If it were the case that p, it would be the case that q iff in all the counterfactually available p worlds, q.

In other words, the role played by "available futures" in my semantics for *will* is played by "counterfactually available worlds" in the relevant semantics for the counterfactual *would*.

But let me back up. There has, of course, been a longstanding debate about the semantics for counterfactuals. At a certain level of description, however, it seems that we can all agree—or most of us can agree—on the following. Whether $p > q$ is true is a matter of whether all of a certain set of p worlds are q worlds. In other words, whether $p > q$ is true is a matter of whether in the relevant p worlds, q.[10] Again, I propose to call the relevant worlds—whatever they are—the "counterfactually available" worlds.

Now, just as, in the case of *will*, we asked which futures are "available", so we can ask which worlds are "counterfactually available". There are two broad traditions of thought in the philosophical literature concerning how we should think about which worlds are counterfactually available. According to one influential tradition, the truth of a counterfactual is always a matter of something *non-modal* (or "categorical"). On this approach, the truth of a counterfactual is a matter of *objective (non-modal) similarity* to the actual world. Roughly, the leading idea here is that $p > q$ is true just in case *all of the most similar p worlds are q worlds.* On this approach, we give what we might call a *reductive* analysis of counterfactuals: the truth of a counterfactual—which is in a certain sense "modal"—always reduces to facts that are themselves *non-modal.* Thus: the counterfactually available worlds are those that are, in the relevant way, objectively most similar to the actual world. Call this sort of non-modal, objective similarity "closeness". To assume that all the closest p worlds are q worlds or all the closest p worlds are $\sim q$ worlds amounts to assuming (local) determinism. Under this

[10] Those who agree that the counterfactual has truth conditions will seemingly agree with this much. Of course, proponents of Edgington-style views (on which they do not) will not. For more on such views, see, e.g., Edgington 1995.

assumption, our ignorance, say, concerning who would have won the game had Jones made the catch is simply ignorance concerning which outcome would have been determined to happen.

Against this approach, however, there has been another. According to some theorists, the truth of certain counterfactuals *cannot* be reduced to facts that are themselves non-modal. Instead, to account for the truth of certain counterfactuals, we must suppose that there are facts that are themselves *primitively modal.* For instance, many people would be happy to grant that fair coins are genuinely indeterministic. And yet they still seem to find room to wonder: how would the coin have landed had it been flipped? On this approach, it seems, even if indeterminism is true, there is nevertheless a primitive "modal hand" that tips the counterfactual scales one way or the other. Consider again the perspective of the "Molinists". Suppose Curley has indeterministic ("libertarian") freedom. And suppose he was never offered the bribe. Well, would he have taken it? According to the Molinist, either if Curley were offered the bribe, he would have taken it, or if Curley had been offered the bribe, he would have rejected it. Crucially, however, given the assumption of Curley's indeterministic freedom, we can suppose that worlds where Curley accepts the bribe and worlds where Curley rejects it are *just as similar* to the actual world. Nevertheless, there is a primitive modal hand, and God knows what it has written, even if we do not.

On this sort of picture, which worlds are counterfactually available is not just a matter of closeness. Instead, there are primitive modal facts—the "counterfacts"—that "break ties" amongst worlds otherwise tied for being objectively similar to the actual world (cf. Hawthorne 2005b and Schulz 2014).[11] That is, even if some *p*

[11] Let me try to clear up a terminological issue about "closeness" and "similarity". As I am using these terms, they are synonyms, and "closeness" and "similarity" are always determined *apart* from any primitively modal counterfacts that may exist. It is worth noting, however, that the terminology of "closeness" is not always used in this way. For instance, Hawthorne writes:

> In closing I might mention that my own preference is to opt for a picture according to which, for any possibility that P, and any world w, there is a unique closest world to w where P. I realize, of course, that this is to give up altogether on the Lewisian idea of analyzing counterfactual closeness in terms of similarity. (2005b: 404)

For me, however, this is akin to giving up on the idea of analyzing closeness in terms of closeness. Of course, this issue is merely terminological; in my framework, we simply say that what Hawthorne is giving up on is the idea of analyzing counterfactual *availability*—i.e., which worlds are counterfactually relevant—in terms of similarity. And this is fine: indeed, precisely Hawthorne's approach is to grant that there may be ties in terms of (objective, non-modal) similarity, but to insist that there are further facts which may be brought to bear which break those ties.

The general lesson here is this: someone who posits primitive modal counterfacts can accept what we might call a "similarity" semantics for counterfactuals, so long as the similarity metric at issue is appropriately sensitive to those primitive counterfacts. There is nothing in principle wrong with talking in this way; however, I believe it is theoretically more perspicuous to reserve talk of "similarity" and "closeness" to similarity and closeness *apart from any primitive counterfactual facts* that may be taken to exist.

This issue is also important to keep straight when considering the views of certain Molinists. Thus, Plantinga, for instance, claims to *accept* a "similarity" analysis of counterfactuals. However, Plantinga then insists that two worlds' "sharing their counterfactuals" must be taken into account when assessing worlds where q and some p worlds where $\sim q$ are just as similar to the actual world, there is also a further consideration, in virtue of which one of these worlds is primitively "selected", so therefore uniquely counterfactually available. Call these primitive modal facts that select one world amongst the most similar the "counterfacts". Consider Stefánsson's recent forthright postulation of such facts:

> I contend, either the coin would have landed heads if tossed or it would not have landed heads if tossed. But if the coin is unbiased and the toss is truly chancy, then there is no ordinary, non-modal fact that determines which way it would have landed. Instead, I suggest, there is a primitive counterfact, that is not entailed by (nor supervenes on) the ordinary facts, but is part of the fundamental structure of reality. (2018: 883)

Here, of course, we have a natural comparison with the *primitive future directed facts* of Chapter 2.

Now the crucial point. With these notions in play, we can give a neutral characterization of "counterfactual availability" as follows: the counterfactually available worlds are *those closest worlds consistent with the counterfacts (if there are any)*. In other words, just as we gave a neutral characterization of the "available futures" as *those causally possible futures consistent with the primitive future directed facts (if there are any)*, so we give a neutral characterization of "counterfactual availability" in terms of those closest worlds consistent with the counterfacts.

We now have our three models to consider. First, just as, in Chapter 2, I simply assumed indeterminism, so I wish to make the parallel assumption that *there can be genuine ties in closeness*. In other words, the three parallel models we must consider here are models which assume that there can be such ties. Thus, assuming that for a given counterfactual $p > q$, there is a tie in closeness, with some of the closest p worlds featuring q, but some featuring $\sim q$, our further options can be stated as follows:

(i) There are primitive counterfacts, and it is determinate what they are. (These primitive counterfacts select a unique p world, and this unique p

those worlds' similarity! (Plantinga 1974: 178; cf. Flint 1998: 135, who is sympathetic; for discussion, see Mares and Perszyk 2011: 110.) Informally, Plantinga's position is the following. Either if Curley had been offered the bribe, he would have taken it, or if Curley had been offered the bribe, he would have rejected it. One is true, and it is determinate which. Which is it? This much is simply a brute modal fact. Well, suppose it is, in the actual world, (brutely) true that had Curley been offered the bribe, he would have taken it. Now, if that is right, then a different possible world in which that is *also* true is thereby more similar to the actual world than any otherwise exactly similar world in which it isn't. So Plantinga can happily say that the given counterfactual is true iff in all of the most similar worlds in which Curley is offered the bribe, he takes it. If the counterfactual is true, then in all of the most similar worlds to the actual world, it is also true, in which case the biconditional is true. Readers can decide whether this proposal is objectionably unilluminating, or what kind of problem this is for Plantinga if it is.

world is either a q world or is not a q world.) Thus, only one p world is counterfactually available, and it is determinate which world that is.

(ii) There are primitive counterfacts, but it is indeterminate what they are. (These primitive counterfacts select a unique p world, and this unique p world is either a q world or is not a q world.) Thus, only one p world is counterfactually available, but it is indeterminate which world that is.

(iii) There are no primitive counterfacts at all. Thus, p worlds in which q and p worlds in which $\sim q$ are *both* counterfactually available.

Plainly, these models are precisely parallel to the models considered in Chapter 2. And as before, I defend (ACW) and model (iii), which together predict that the relevant counterfactuals are *false*, and thus that CEM is false. Notably, however, (ACW) together with either model (i) or model (ii) *retains* CEM; on these views, since there is always a unique world that is counterfactually available, that unique world will always be a q world or instead a $\sim q$ world. The cost associated with these views is metaphysical or ontological: both are committed to primitive counterfacts. The challenge for (ACW) together with model (iii) is, in some sense, semantic: it must explain or explain away the linguistic data that would seem to support CEM.

To sum up. In the case of counterfactuals, I defend a standard semantics for counterfactuals, maintain that sometimes there can be ties in closeness, and deny on metaphysical grounds that there are, in such cases, primitive counterfacts that break those ties, and thus reach the conclusion that the relevant counterfactuals—"counterfactual contingents", we might say—are false. In the case of *will* claims, I defend a parallel semantics for *will*, maintain that sometimes there are multiple causally possible futures, and deny on metaphysical grounds that there are primitive future directed facts that select one such future, and thus reach the conclusion that future contingents are all false. As I said above: some novelty is good, but not too much. Here it can be seen that, in some crucial respects, my approach to *will* is not novel at all; it mirrors precisely well-known positions already present in the literature on counterfactuals.

4.5 Flow-Charts

Perhaps the easiest way to bring out the close parallels between the way I am treating *will* and *would* is to compare the flow-chart from Chapter 2 to a parallel such chart in the case of *would*[12] (Figure 4.1).

[12] My sense (which I shall not try to justify) is that the views I am associating with Stalnaker-1 and Stalnaker-2 are both sometimes attributed to Stalnaker; at any rate, I set this interpretive issue aside.

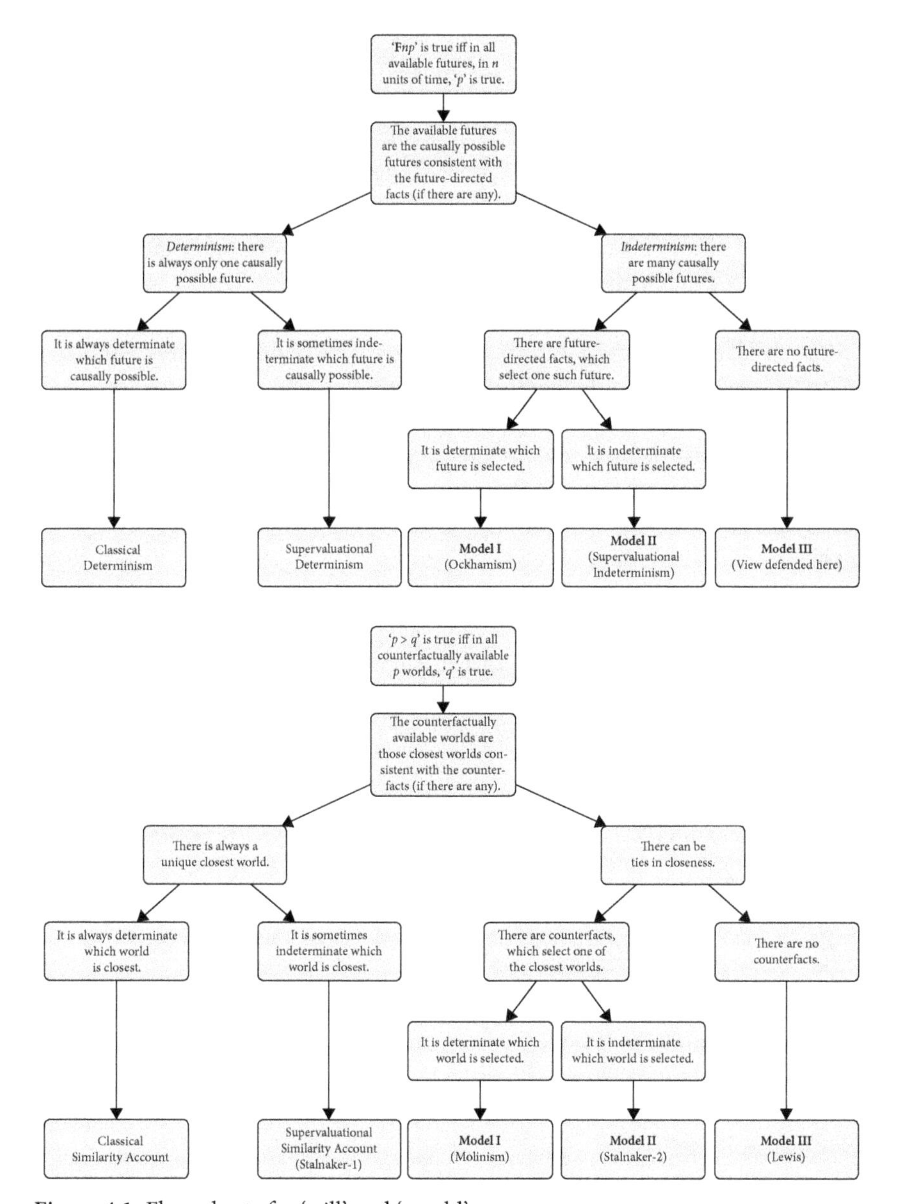

Figure 4.1 Flow-charts for 'will' and 'would'

4.6 The Neg-Raising Inference

Above I provided what I called an unofficial statement of the relevant "neg-raising" inference from $\sim(p > q)$ to $(p > \sim q)$. However, with the above distinctions on the table, we can now state that inference slightly more carefully. On my proposal, the reason we so easily seem to "slide" from ~($p > q$) to (p >~q) is that, unless in a "philosophical" context, we implicitly tend to assume the following:

> (CON) All of the counterfactually available p worlds are q worlds or all of the counterfactually available p worlds are ~q worlds.

Or, more to the point, given the neutral characterization of "counterfactual availability", we implicitly tend to assume:

> (CON*) All of the closest p worlds consistent with the counterfacts are q worlds, or all of the closest p worlds consistent with the counterfacts are ~q worlds.

Now, there might be *two* reasons that we implicitly tend to assume CON*. Perhaps some of us are implicit determinists, who suppose that the facts about closeness *themselves* always select one particular world. And perhaps some of us implicitly think that where those facts don't pick out a unique p world, some further primitive facts nevertheless do. In any case: we implicitly tend to assume CON*. The story from here is then familiar: because we assume CON*, when someone *denies* that ($p > q$)—says that it is false that all of the closest p worlds consistent with the counterfacts are q worlds—we hear that as equivalent to (p > ~q)—i.e., we hear that as an assertion that all of the closest p worlds consistent with the counterfacts are ~q worlds. However: CON* can be coherently denied; it can be denied if there can be ties in objective, non-modal similarity, and there are no counterfacts. If CON* is false, CON is false—and if CON is false, so is CEM.

4.7 On an Argument for CEM from Quantifiers

In the rest of this chapter, I respond to a critical argument for the truth of CEM—and, by extension, WEM.[13] J.R.G. Williams is skeptical of Williamson's arguments against CEM; indeed, Williams articulates a new argument *for* CEM. And here we come, at last, to the parallel scope distinctions defended in Chapter 3 (concerning "No ticket will win"):

> One might consider explaining away the intuitions backing CEM (Williamson 1988). There are, however, reasons for thinking that its role runs deeper than merely the intuitive appeal of this principle. For example, von Fintel and Iatridou

[13] In what follows, I give a condensed version of the reply I have jointly developed with Brian Rabern; see our (forthcoming) paper, "If Counterfactuals Were Neg-Raisers, Conditional Excluded Middle Wouldn't Be Valid".

> (MS 2002) connect CEM with the behavior of conditionals under extensional quantifiers. Their work suggests the argument that follows:
> **Premise 1:** The following are equivalent:
> A. No student would have passed if they had goofed off
> B. Every student would have failed to pass if they had goofed off.
>
> **Premise 2:** (A) and (B) can be regimented respectively as follows:
> A*. [No x: student x](x goofs off → x passes)
> B*. [Every x: student x](x goofs off → ¬ x passes)
>
> **Premise 3:** For any *F*, "[No x: Fx]Gx" is equivalent to "[Every x: Fx] ¬Gx"
> From these three premises, we argue as follows. By an instance of premise 3, A* is equivalent to:
> C*. [Every x: student x] not(x goofs off → x passes)
> There is then a chain of equivalences running from (C*) back to (B*). (C*) is equivalent to (A*), which is equivalent to (A) (premise 2) which is equivalent to (B) (premise 1) which is equivalent to (B*) (premise 2). So (C*) is equivalent to (B*).
>
> It seems to me that we can generalize each of 1–3. *Quite generally*, 'No *F* would *G* if they *H*' and 'Every *F* would fail to *G* if they *H*' seem equivalent...
>
> So, from the intuitive equivalence of the embedded counterfactuals in premise (1), we derive instances of the equivalence of negated and opposite conditionals. The corresponding instances of CEM follow classically in a couple of steps.... The upshot is that denying CEM comes to seem rather heroic: one needs to explain away, not only the intuitive appeal of $[(p > {\sim}q) \leftrightarrow {\sim}(p > q)]$, but also the apparent equivalences between quantified conditional statements.[14]

We can be heroes. Williams' argument, however, is a powerful one—and, in my judgment, it takes some substantial groundwork to see how it fails. Notably, if Williams' argument works in the case of *would*, a parallel argument should work in the case of *will*. It is thus crucial for my purposes to see how Williams' argument fails. I thus pursue the following roundabout strategy: I first develop a parallel response in the case *should*. We can then apply this strategy to the case of *would*—and then finally to *will*.

4.8 Should

The first thing to say about this argument is that, although it is a delicate matter explaining exactly *how* this argument goes wrong, there is nevertheless substantial

[14] Williams 2010: 652. A version of this argument is likewise developed by Mandelkern 2018 (who attributes it in turn to Higginbotham 1986 and 2003); it is similarly endorsed in Goodman ms.

reason to believe that it *does* go wrong. For the above argument can be parodied to generate an argument for the unwelcome principle of *Should Excluded Middle*. Recall: the above argument for CEM proceeds solely on the basis of an intuitive equivalence between claims like:

(1) No student would pass were he to goof off.
(2) Every student would fail to pass were he to goof off.

(3) No employee would be happy were she to be sacked.
(4) Every employee would fail to be happy were she to be sacked.

And so on. And the fact is that these claims *do* sound equivalent (at any rate, they certainly sound like claims only philosophers or linguists would try to prize apart). The problem for this argument, however, is that a parallel observation holds for *should*:

(5) No student should pick up a hitchhiker if he is traveling alone.
(6) Every student should fail to pick up a hitchhiker if he is traveling alone.

(7) No employee should report to work if she has a fever.
(8) Every employee should fail to report to work if she has a fever.

Examples could be multiplied, but the point is clear. *Quite generally*, 'No *F* should *G* if they *H*' and 'Every *F* should fail to *G* if they *H*' seem equivalent. And yet: can we use this intuitive equivalence to generate an argument for some principle like Should Excluded Middle (SEM) viz., 'It should be that *p* or it should be that ~*p*', or perhaps 'For all *S*, *S* should *phi* or *S* should not *phi*'?

My claim here is simple. My claim is that once proponents of the above argument for CEM explain where something goes wrong with the relevant argument for SEM, we will be able to use a parallel strategy to explain what goes wrong with the original argument for CEM. Thus: how should we explain why (5) and (7) *seem* equivalent to (6) and (8)—although in fact they aren't? And at least one answer is as follows: by remembering that *should* is a neg-raising predicate—and remembering that, in order to trigger the relevant readings of such predicates, we will have to place emphasis in the right place. Recall the example (slightly modified) from the end of Chapter 1 (that is, recall that *wants* is a neg-raiser):

Jack, no one here wants to go to your stupid party.

Pragmatic Implication: Everyone here wants to not go to your party.

Jack, no one here *wants* to go your stupid party—then again, no one here wants to *not* go to your stupid party; everyone here is perfectly neutral about going to your stupid party. (– said the smart-aleck)

Similarly, by emphasizing the *should* in the above, we can achieve a reading of (5) that does not entail (7):

> Our students certainly don't have any special reason or obligation to pick up hitchhikers—especially when traveling alone. So no student *should* pick up a hitchhiker when traveling alone—but then again, I'm not saying that every student should *refrain* from picking them up; the area is safe, so although students need not pick up hitchhikers, they need not *not* pick them up either.

The relevant reading of "No student *should*..." is one on which we are saying that no student is *such that* he should pick up hitchhikers when traveling alone. But it is relatively clear that it does not follow from this fact that every student is such that he should *refrain* from doing so: for perhaps the students have no obligations in this matter either way.

And now perhaps you see where we're headed. The question is whether we can achieve a similar reading of "No student *would* have passed were he to have goofed off". If we can, then that reading is one on which we are saying that no student is *such that* he would have passed were he to have goofed off. But now the point is clear: the denier of CEM thinks it is coherent, on metaphysical grounds, to suppose that it could be that no student is such that he would have *passed*, but *also* no student is such that he would have *failed*. The point is perhaps easier to observe for indeterministic coins. Consider:

> (9) No indeterministic coin is such that it would have landed *heads* had you flipped it, but not every indeterministic coin is thereby such that it would have landed *tails* had you flipped it.

I am not asking the reader to judge that this statement is perfectly acceptable; I am instead asking the reader to observe that it appears simply to be a statement of the CEM-denier's position. Certainly the *proponent* of CEM cannot offer the obvious unacceptability of (9) as *evidence* against the denial of CEM; for that is tantamount to offering a statement of the denial of CEM as evidence against the denial of CEM. Thus, the crucial question is whether we can achieve the relevant readings of (1) and (3)—in which case, the denier of CEM will be within his or her rights (well, whatever rights he or she had already) in maintaining that these are not equivalent to (2) and (4).

(1) No student would pass were he to goof off.
read as: No student is such that he would pass were he to goof off. *Does not entail:*
(2) Every student would fail to pass were he to goof off.

Just as in the case of *should*, the relevant reading can be made salient by emphasizing the *would*: "No student *would* pass were he to goof off" *can* mean, in a suitable context, that no student has the property of being such that he would pass were he to goof off. And in that context, to insist that this claim is equivalent to the claim that every student has the *opposite* property (being such that he would *fail* to pass were he to goof off) is simply to beg the question against the denier of CEM. Plainly, we don't think that *being such that you would pass* or *being such that you would fail* exhaust the options; there is also a third: being neither.[15]

So ends my reply to Williams: *contra* his first premise, the relevant claims are not in fact equivalent, and the argument for CEM fails. And now we can come full circle, and apply this lesson to *will*. Plausibly, Williams could have run a closely parallel argument for Will Excluded Middle, one moving from the seeming equivalence of "No ticket will win" and "Every ticket will lose". But as we have seen, the fact that these claims *seem* equivalent at first blush can be adequately explained by the thesis that *will*—like *should*—is a neg-raising predicate. Thus, we have found a way of saying, what is no doubt a strange, awkward thing to say: "None of these options *will* obtain, but it will be that one obtains."[16]

4.9 Conclusion

But let me now back up. My point in this chapter, again, is not so much to argue against CEM. It is instead to point out that there is a perfectly respectable, common philosophical position that *requires* us to argue in these ways against CEM. That is, deniers of CEM—and there are many—must find a way of maintaining that "No ticket would have won if we had run the lottery" does not unambiguously entail "Every ticket would have failed to win if we had run the lottery." But then we can use the *exact same tools* to explain why "No ticket will win" does not unambiguously entail "Every ticket will lose." (Recall this discussion in the previous chapter.) More generally, whatever resources we use to maintain

[15] Objection: this *would* has now become *would definitely*. Response: I don't see why this is so, any more than I see why *want* had become *definitely want* as discussed above (Chapter 3)—or any more than *should* has become *should definitely*, when Jones says, "No ticket here *should* win." Jones isn't saying that, though some ticket here does have the property of being such that it should win, there is just no ticket such that it *definitely* or *determinately* has this property. He just thinks, as we said, that no ticket has this property at all. So similarly in the cases of *would* and *will*.

[16] Of course, though this claim is strange and awkward, *philosophers* do sometimes make (or defend the felicity of) claims of precisely this form; cf. this apt comment from Mitchell Green:

> Rather, it is compatible with Open Future to hold that such branches represent, "ways history might go." Standing at an indeterministic point, then, we might say of each of the possible future courses of events, "This is a way that history might go; all we claim now is that none of these is what *will* happen." (2014: 157)

Green, of course, is not endorsing the obviously false claim that it will be that none of these futures happen. He is instead endorsing the claim that no one history is, now, such that *it* will happen.

the important scope distinction between ~($p > q$) and ($p > {\sim}q$) we can *also* use to defend the important scope distinction between ~Fnp and Fn~p. And this is what matters. If you were skeptical of those distinctions in Chapter 3, then you must be prepared to claim the CEM is a *semantic* truth. And those who criticize my deployment of such distinctions cannot single me out for selective opprobrium: any denier of CEM, e.g., Williamson and Lewis, must make parallel distinctions in the case of *would*.

If you think that counterfactuals must be grounded in the categorical, then an attractive option for you is to deny CEM and maintain that both of the relevant counterfactuals are *false*. Similarly, if you think that future-tense truths must be grounded in the *present*, then an equally attractive option is to deny Will Excluded Middle, and say that future contingents are false. *Semantically*, it seems, both of these options are on a par. *Metaphysically*, however, it may be plausible to suppose that future-tense claims do *not* need to be grounded in the present, whereas counterfactuals *do* need to be grounded in the categorical. But this is a different matter: once the metaphysical claims are granted, the semantic claims stand or fall together.

5

Omniscience and the Future

My aim in this chapter is to display how my own account of the open future allows us a simple, attractive theory of omniscience. At least, *if* we are persuaded to give up this account of omniscience, it will be for reasons *other* than the openness of the future. I defend a strong conception of omniscience that makes *p* logically equivalent to *God believes p*. Once again, my point in this chapter is not so much that this equivalence in fact holds (and that therefore there exists an omniscient being), but that considerations arising from the *open future* give us no reason to think it doesn't. My second aim is more preparatory. In the chapters to come, I invoke at various points the intuitive equivalence, defended below, between *there will be a sea-battle tomorrow* and *an omniscient being* [God] *anticipates a sea-battle tomorrow*. In this chapter, I introduce and motivate this intuitive equivalence.

5.1 Two Versions of Open Theism: Open Future, and Limited Foreknowledge

When combined with traditional theism, the "open future" view I defend in this book gives rise to a version of what has come to be called "open theism". Roughly, the open theist rejects the classical picture of divine foreknowledge, and instead accepts the view that, for some relevant events, God knows neither that they *will* occur nor that they will *not* occur.[1] In this, sense, the future is open—even for God. Now, there are two importantly different versions of open theism: what we might call *Open Future Open Theism* (the view currently defended) and *Limited Foreknowledge Open Theism*.[2] In contrast to the open futurist, the Limited Foreknowledge Open Theist is an *Ockhamist* when it comes to the logic of future contingents; on this view, some future contingents are indeed just plain true. The crucial contention of the Limited Foreknowledge Open Theist is that *not even God* could know these truths; more particularly, whereas on this view there are facts about how indeterministic processes shall unfold, God cannot have, in advance, access to these facts. The result is that though the future is not "open" with respect

[1] Rhoda 2008; Rhoda provides a helpful general characterization of "open theism" along the lines I provide here.

[2] Todd 2014.

The Open Future: Why Future Contingents are All False. Patrick Todd, Oxford University Press. © Patrick Todd 2021.
DOI: 10.1093/oso/9780192897916.003.0006

to *truth* (there is indeed a "privileged future history" and a complete "story of the future") it is nevertheless epistemically open for God.

Open theism is sometimes described as a view that must either deny or qualify the doctrine of God's omniscience. But this is a mistake. Only the *Limited Foreknowledge* open theist must deny or qualify the claim that God is omniscient; the open futurist, however, can maintain the plausible claim that the *facts* about the future do not outstrip what a perfect God *knows* about the future. Indeed, at this stage, it is worth reviewing what many will see as a core metaphysical advantage of the doctrine of the open future. For consider the claim—made by both versions of open theism—that not even God could know, in advance, the outcomes of genuinely indeterministic processes. It has long been recognized even by proponents of God's foreknowledge of such facts that such foreknowledge must inevitably strike us as deeply mysterious—if not mystical.[3] For suppose it were a true future contingent—and known to God—that there will be outposts on Mars 1000 years from today. Ex hypothesi, since the given claim is a future *contingent*, *nothing at all* about the current arrangement of things—and nothing at all about God's own plans or intentions—fixes it that in 1000 years there will be such outposts. So how then is God achieving such foreknowledge? By being outside of time? Then the knowledge isn't *foreknowledge*—and, at any rate, we are assuming that God is not "outside of time".[4] By means of a crystal ball? But that is just repeating our mystery. By "seeing" into the future? But that is incoherent (and at any rate surely anthropomorphic). How then? Well, we don't know.[5]

5.2 The Logic of Omniscience

The primary goal of this chapter, however, is not to provide further metaphysical argumentation for the doctrine of the open future; it is instead to provide at least some philosophical motivation for my particular *version* of that doctrine—viz., one according to which future contingents are systematically false. Here I turn, then, to what we might call the logic of temporal omniscience. I will not offer a full account of omniscience; instead, I will begin with a plausible *constraint* on omniscience that I will call *omni-accuracy*:

Omni-accuracy. p iff God believes p. (p iff $\mathbf{Bel}p$)

[3] Hence Ockham's famous admission: "It is impossible to express clearly the manner in which God knows future contingents" (Ockham in Adams and Kretzman 1969: 50).

[4] The doctrine of the open future would seem to sit badly with a theistic picture on which God is outside of time (Stump and Kretzmann 1981, Leftow 1991)—but I set these points aside.

[5] However, Byerly (2014) suggests some "conciliatory stories" meant to make this result more palatable. Cf. Plantinga 1993.

I will begin with the assumption that omniscience entails omni-accuracy. (Later we will consider a different characterization of omniscience that relaxes this assumption.) It follows from omni-accuracy that

Pn*p*/**F**n*p* iff God believes **P**n*p*/**F**n*p*.

For ease of exposition, I will often talk in terms of God's *recollections* and God's *anticipations*. That is, in the case of God, to believe that something was the case is to remember its being the case; thus, for sake of simplicity, we may write the following:

It was n units of time ago that *p* iff God remembers than n units of time ago, *p*. (**P**n*p* iff **Rem**n*p*)

And, in the case of God, to believe that something will be the case is to anticipate its being the case:

It will be in n units of time hence that *p* iff God anticipates that in n units of time hence, *p*. (**F**n*p* iff **Ant**n*p*)

Accordingly, as I hope to show, my view can happily accept—whereas other open future views cannot—the plausible thesis that *the logic of tense is the logic of perfect memory and anticipation*. The theory defended in this book is that, ignoring so-called "hyper-intensional" contexts,[6] **F** and **Ant** (and **P** and **Rem**) are inter-substitutable: whenever we see **F**, we can replace it with **Ant**, and vice versa, and preserve truth. As we'll see, I believe that theorizing in terms of these equivalencies provides a helpful framework for investigating certain crucial principles of tense-logic.

5.3 Perfect Anticipation: Variations on a Priorean Theme

I turn first to the question whether the intuitive equivalence between *There will be a sea-battle tomorrow* and *God anticipates a sea-battle tomorrow* provides any motivation for seeing (or perhaps reluctantly accepting) the scope distinctions I have earlier defended in this book—in particular, the crucial distinction between ~**F**n*p* and **F**n~*p*, and the associated denial of "Will Excluded Middle" (WEM). Now, it is clear that, when it comes to *will*, it can seem that there are only two options:

[6] For example, embedded contexts such as *belief*. The fact that Jones believes **F**n*p* certainly doesn't imply that Jones believes **Ant**n*p*. The idea I defend is instead that *in any principle of tense-logic*, **F** and **Ant** can be substituted while preserving truth.

there will be a sea-battle tomorrow or there will be no sea-battle tomorrow. But the relevant equivalence, intuitively, opens up the conceptual space to see a third. For if *it will be that there is a sea-battle tomorrow* is equivalent to *God anticipates a sea-battle tomorrow*, then *it will be that there is no sea-battle tomorrow* is equivalent to *God anticipates an absence of a sea-battle tomorrow*. But now the point. Intuitively, *if the future is genuinely open, then God may as of yet have no anticipation either way*. In that case: we have neither **Ant**n*p* nor **Ant**n~*p*—or, more to the point, we have both ~**Ant**n*p* and ~**Ant**n~*p*. But if **Ant** and **F** are equivalent, then if the future is genuinely open, we have both ~**F**n*p* and ~**F**n~*p*. And there we are. If the future is open, future contingents are all false, and WEM is denied.

Consider the following:

G1. God anticipates (its becoming the case that) *p* in n units of time. (**Ant**n*p*.)
G2. God does not anticipate (its becoming the case that) *p* in n units of time. (~**Ant**n*p*.)
G3. God anticipates (its becoming the case that) ~*p* in n units of time. (**Ant**n~*p*.)

We can now ask: does G2 entail G3? (Note: in what follows, I sometimes simplify by omitting the 'its becoming the case that' locution, writing simply [if a bit inaccurately] 'God anticipates *p*'.)

Certainly many have thought that it does; Ockham himself (after whom *Ockhamism)* would certainly contend that it does. That is, on the standard, classical picture of divine omniscience, defended at length by Ockham, if God *does not* anticipate a sea-battle tomorrow, that is going to be *because* God anticipates the *absence* of a sea-battle tomorrow. And this is plainly because, for Ockham, God, being God, somehow has access to which future history is the privileged future history—that is, the facts about how indeterministic processes shall unfold (if there are any such processes). Thus, for a theorist like Ockham, we might say the following: For God, *absence of an anticipation is always anticipation of an absence*. If, in God, there is no anticipation of *p* in n units of time, and so G2 is true, this is going to entail that, in God, there is an anticipation of ~*p* in n units of time. Thus, for a theorist like Ockham, if God does not now anticipate a sea-battle tomorrow, then you can rest easy: this by itself implies that there is going to be no sea-battle tomorrow.

But now the crucial point: though G2 may indeed entail G3, it is obvious that this entailment is *substantive*, not semantic. That is, it is obvious that G2 cannot simply be *rewritten* as G3—as if G3 carries with it no new semantic content over and above G2. The argument that G2 does indeed entail G3 takes us well beyond mere semantic competence. Indeed, my own picture of the open future might be perfectly captured in the following slogan:

> *Even for an omniscient being, absence of anticipation does not always imply anticipation of absence.*

More particularly, we suppose that God is perfectly omniscient regarding the future. We then ask God:

US: Do you anticipate a sea-battle tomorrow?
GOD: No. (~**Antn***p*)
US: Do you anticipate there *not* being a sea-battle tomorrow?
GOD: No. (~**Antn**~*p*)
US: So you're saying the future is open?
GOD: Precisely. I don't anticipate a sea-battle tomorrow, but then again, I don't anticipate the *absence* of a sea-battle tomorrow. At this stage, I have no anticipation either way. (~**Antn***p* ∧ ~**Antn**~*p*)

If the future is genuinely open, in the sense I wish to defend in this book, then from the fact that an omniscient being does not have an anticipation of its becoming the case that *p* in n units of time, it simply would not follow that that being has an anticipation of its becoming the case that ~*p* in n units of time. That is, for some *p*, it would be false that **Antn***p*, and false that **Antn**~*p*. And so similarly for the propositions to which these seem equivalent: Fn*p* and Fn~*p*.

In sum, G1–G3 seem perfectly equivalent to W1–W3.

G1. God anticipates *p* in n units of time.
G2. God does not anticipate *p* in n units of time.
G3. God anticipates ~*p* in n units of time.
W1. It will be that *p* in n units of time.
W2. It is not the case that it will be that *p* in n units of time.
W3. It will be that ~*p* in n units of time.

And if these are equivalent, then if G2 does not semantically entail G3 (and it doesn't), then W2 should not semantically entail W3. G1 and G3 are semantic contraries. At any rate, if both can't be false, this is one's *theory* talking, not one's semantic competence. So similarly for W1 and W3.

5.4 God's Tickets

But now recall the problems from Chapters 3 and 4 about *quantifiers*. In particular, recall that scopelessness predicts, whereas my own theory denies, that the following are truth-conditionally equivalent (assuming, once more, a constant domain of objects over time):[7]

[7] Note: for ease of reference, I sometimes write the longer '**Will**' in place of '**F**'.

It will be that there is a winning ticket.
Will: $\exists x$, x a ticket, x a winner
There is a ticket such that it will be the winning ticket.
$\exists x$, x a ticket, **Will**: x a winner

The current theory of omniscience has it that **F** and **Ant** are intersubstitutable. Hence, if this is accepted, then scopelessness predicts that the following are truth-conditionally equivalent, whereas my view predicts that they aren't:

Ant: $\exists x$, x a ticket, x a winner
$\exists x$: x a ticket, **Ant**: x a winner[8]

In particular, proponents of scopelessness will have to endorse the following:

> God anticipates there being a ticket that is the winning ticket iff there is a ticket such that God anticipates of it that it will be the winning ticket.

Whereas everyone can admit that *someone* (a mere mortal) might anticipate that it will be that some ticket wins, *without* anticipating of any particular ticket that *it* will win, the proponent of scopelessness must say: yes, *but not if you are omniscient.* If you are omniscient, then if you anticipate that it will be that some ticket wins, then it will be that some ticket wins—and if it will be that some ticket wins, then some ticket is such that it will win—and if some ticket is such that it will win, and you don't believe of any ticket that it will win, then, well, you aren't omniscient. For then we would have:

$\exists x$, x a ticket, **Will**: x a winner $\wedge$ $\sim\exists x$, x a ticket, **Ant**: x a winner

In which case **F** and **Ant** are not equivalent, in which case God is not omniscient.

I think there is a problem with this argument. The scopelessness theorist is saying that if it will be that some ticket wins, unless you believe of some ticket that *it* will win, *you are thereby ignorant*. But is this so? Of course, on *Ockhamist* assumptions, this implication is clear. (If some ticket will win, there is a fact of the matter about which.) But why should it hold on *open-futurist* assumptions? In other words, suppose we grant that it is *determined* that it will be, later, that some ticket wins the lottery; on every branch, some ticket or other (of tickets 1, 2, and 3) wins. On open-futurist assumptions, however, it could nevertheless be that there is *no fact of the matter* about *which* ticket will win. But if there is no fact of the matter about which ticket will win, why should it follow that you are *ignorant* if you do

[8] Again, as in Chapter 4, I bypass the issue of whether the tickets *exist* yet, and assume a constant domain of objects over time.

not believe of any particular ticket that *it* will win? Intuitively, this shouldn't follow. But scopelessness predicts that it does.

Consider the following. If there is no fact of the matter about which ticket will win, what should we expect God to *say* if we asked of some given ticket whether he anticipated its winning? Intuitively, we should expect God to say 'No'. For instance, we ask God of ticket 1: do you believe that it will win? God says 'No'. Ticket 2? No. Ticket 3? No. But we ask: do you believe that it will be that some ticket or other wins? And of course here God says 'Yes'; this much, we said, is determined. Scopelessness says: straightaway, God is ignorant. I say: no, God can still be omniscient, because whereas we have '**Will**: ∃x, x a ticket, x a winner'—and this much God certainly believes—we do *not* have '∃x, x a ticket, **Will**: x a winner'. God rightly anticipates that it will be that some ticket wins, but rightly does not anticipate of any particular ticket that *it* will win—and, I say, God remains omniscient, because there is no fact of the matter concerning which ticket will win. As I see it, this is just as it should be.

This response, however, is not open to the (open-futurist) proponent of scopelessness/Will Excluded Middle. But then what does this proponent of scopelessness *want* God to say, when we ask him the relevant questions? When we ask God whether he believes of the given ticket that it will win, apparently God can't just say *no*. And, evidently, God can't also at any point simply say *yes*; that is the response of the Ockhamist. But then what *does* God say? Presumably:

US: Do you believe of Ticket 1 it that it will win?
GOD: That's indeterminate.
US: Ticket 2?
GOD: That's indeterminate.
US: Ticket 3?
GOD: That's indeterminate.
US: And those are the only tickets?
GOD: Yes.
US: So for each ticket, it is indeterminate whether you anticipate it being the winner.
GOD: Yes.
US: But some ticket is such that you anticipate of it that it shall win.
GOD: Yes. It is just indeterminate which ticket that is.

But no. "God believes of one of these tickets that it will win, but it is just indeterminate which ticket that is" does not have the ring of truth. It is far more natural to say that, if there is no fact of the matter about which ticket will win, though God anticipates that it will be that one of these tickets wins, there is no ticket such that God believes of *it* that it will win. But what we say about **Ant** we should say about **F**. So we should say that though it will be that one of these tickets is the winner, there is no ticket such that *it* will be the winner. And if we say *that*, WEM and scopelessness are false.[9]

The result here is the following. On pain of an implausible sort of indeterminacy in the divine mind, any open future theory on which *will* is scopeless (and WEM is preserved) must deny that **F** and **Ant** are intersubstitutable. They must deny that God believes *p* iff *p*. If what it is for a being to be omniscient, however, is at least in part for that being to satisfy the schema (*p* iff **Bel***p*), then scopelessness (together with the open future) predicts that there is no omniscient being. And isn't that odd? For my own part, I am inclined to say that since there *could* be an omniscient being, even if the future is open, we should reject scopelessness, and maintain that there is an important difference between ~Fn*p* and Fn~*p*.

5.5 From Omni-Accuracy to Omni-Correctness

We have started with a conception of omniscience on which omniscience entails

Omni-accuracy: *p* iff **Bel***p*.

And we have brought out how if we accept WEM, but maintain that (in the relevant case) neither disjunct is true, the result is indeterminacy in the divine mind. Such a theory must make sense of the problematic claim that God's mind is, in some respects, *indeterminate*.[10] My claim is not that no sense can be made of this claim. My claim is that the present theory has no need to make sense of it, and that this is a substantial advantage for the present theory.

If we reject this equivalence, however, what is our theory of omniscience? Start with the following observation. For the theory in question, we have Fn*p* ∨ Fn~*p*, but we also have it that neither such claim is *true*—a point we might write as follows, letting 'T' stand for 'It is true that...': ~TFn*p* and ~TFn~*p*. Thus, whereas we have the disjunction of Fn*p* and Fn~*p*, we do not have the parallel disjunction of TFn*p* and TFn~*p*. Thus, a theory on which WEM is true but neither disjunct is true is committed to the non-equivalence of *p* and *it is true that p*. Given this result, the more promising conception of omniscience for this theory does not maintain that (*p* iff **Bel***p*), but instead connects God's beliefs to what is *true*. Call such a principle *omni-correctness*:

[9] As explained in Chapter 3, the problem here is a variant on the following. For the proponent of scopelessness, Fn(1 ∨ 2 ∨ 3) entails (Fn1 ∨ Fn2 ∨ Fn3). Thus, given omni-accuracy, **Ant**n(1 ∨ 2 ∨ 3) should entail (**Ant**n1 ∨ **Ant**n2 ∨ **Ant**3). However, it is, I suggest, deeply implausible, on open-futurist assumptions, that **Ant**n(*p* ∨ *q*) should entail (**Ant**n*p* ∨ **Ant**n*q*). Since **Ant** is equivalent to **F**, I contend that it is likewise implausible, on open-futurist assumptions, that Fn(*p* ∨ *q*) should entail (Fn*p* ∨ Fn*q*).

[10] My co-author and I discuss this possibility in greater detail in Chapter 8.

Omni-correctness: $\mathrm{T}p$ iff $\mathbf{Bel}\mathrm{p}$.

On this open future view, we retain WEM ($\mathbf{Fn}p \vee \mathbf{Fn}{\sim}p$), but we say that (in the relevant case) neither disjunct is true, although the disjunction is true. However, since ${\sim}\mathbf{TFn}p$ and ${\sim}\mathbf{TFn}{\sim}p$ (the future is open!), by omni-correctness we have ${\sim}\mathbf{Antn}p$ and ${\sim}\mathbf{Antn}{\sim}p$. God does not anticipate a sea-battle tomorrow, and does not anticipate the absence of a sea-battle tomorrow—but that is because it is it not *true* that there will be a sea-battle tomorrow, and it is not true that there will be *no* sea-battle tomorrow—although, of course, on this view, there will be a sea-battle tomorrow or there will be no sea-battle tomorrow. So far so good. The problem comes once we combine WEM with the claims about God's anticipations. Once we do so, we get the following:

$\mathbf{Fn}p \wedge {\sim}\mathbf{Antn}p \vee \mathbf{Fn}{\sim}p \wedge {\sim}\mathbf{Antn}{\sim}p$

There will be a sea-battle tomorrow and God doesn't anticipate a sea-battle tomorrow *or* there will be no sea-battle tomorrow and God doesn't anticipate there being no sea-battle tomorrow.

Assuming WEM, this is a result open-futurist proponents of omni-correctness must simply accept. And yet this *sounds* like saying that, either way, God is ignorant. However, it should not be a consequence of the view that the future is *open* that God is ignorant. For this reason, the open future combined with omni-correctness faces at least one seriously counterintuitive result.

The proponents of this combination of views can give the following response: although the above disjunction does *sound* like saying that God is not omniscient, it does not in fact imply that God is not omniscient—for, on this view, there is no *truth* of which God is ignorant, since although the disjunction is true (and God of course believes this disjunction), neither disjunct is true. Formally, this result is exactly parallel to the following (which the given view must also accept): "There will be a sea-battle tomorrow and it isn't true that there will be a sea-battle tomorrow or there will be no sea-battle tomorrow and it isn't true that there will be no sea-battle tomorrow."[11] If proponents of WEM and the open future have learned (or can learn) to live with *that*, then perhaps they can also learn to live with the above. Happily, the current theory needs to learn to live with neither.

There is no decisive argument that it is unacceptable to reject God's omni-accuracy, and instead accept only God's omni-correctness. And there is no decisive argument that the relevant result for the latter view ($\mathbf{Fn}p \wedge {\sim}\mathbf{Antn}p \vee \mathbf{Fn}{\sim}p \wedge {\sim}\mathbf{Antn}{\sim}p$) is simply unacceptable, on assumption that God is omniscient (believes all and only the *truths*). My central point in this chapter thus far is merely

[11] Cf. Hawthorne (in Chapter 7).

the following. On my view, if we accept omni-accuracy—the initially plausible, default conception of omniscience—we do not get any implausible indeterminacy in God's mind: God simply does not believe that there will be a sea-battle tomorrow, does not believe that there will be no sea-battle tomorrow, but believes that it will be tomorrow that there is or is not a sea-battle.

By way of summary, it may help to state the core argument developed above slightly more formally. Consider:

1. It is not true that there will be a sea-battle tomorrow and not true that there will be no sea-battle tomorrow. (Premise: the open future)
2. If *p* is not true, it is false that God believes that *p*. So,
3. It is false that God believes that there will be a sea-battle tomorrow, and false that God believes that there will be no sea-battle tomorrow.
4. God believes *p* iff *p*. (Premise: omni-accuracy)
5. If (*p* iff *q*), then if *p* is false, *q* is false. So,
6. It is false that there will be a sea-battle tomorrow and false that there will be no sea-battle tomorrow.

(1) is our starting point, and (5) is something I shall hereby assume. That leaves (2) or (4). We have considered a response that denies (2) and retains (4): on this picture, sometimes when a claim isn't true, it also isn't *false*—and in that scenario, it isn't *false* that God believes this claim, but instead *indeterminate*. And the cost for this position is that it must make sense of the problematic idea that God's mind might be, in this way, indeterminate. We have also considered a response that denies (4) and retains (2): on this picture, considerations of the open future would show that there is no omni-accurate being. However, for my own part, I consider (2) to be deeply plausible, and I consider (4) to be plausible as well; at least, I do not see compelling reason from considerations of the open future to deny the truth of (4). Thankfully, we can simply accept (6).

5.6 Shifting Gears

It is at this stage in the book that I transition from a development of my view on which future contingents are all false to a series of what we might call *practical* problems for the doctrine of the open future. And, when confronting these problems, somehow it seems to help to recall that the propositions I am saying are all false are equivalent to propositions about God's current anticipations. It is, I contend, easier to see how we can make do with the falsity of such claims about God's anticipations than it is to see, directly, how we can make do with the falsity of future contingents. For instance, consider the following. Suppose we are absolutely sure that there exists an omniscient being—God. To say that it is false that there will be a sea-battle tomorrow is now to say nothing more nor less than that God does not anticipate a sea-battle tomorrow. An omniscient being doesn't anticipate a sea-battle tomorrow. Well, so? Perhaps that being doesn't anticipate an *absence* of a sea-battle tomorrow. As we'll see, if we can make this appeal, then certain problems for the open futurist become easier to address. Further, as we shall see in the chapters to come, the open futurist has long been vexed by a principle I will call *Retro-closure:*

$p \rightarrow \mathbf{PnFn}p$

But now note that, on my view, this principle is equivalent to the principle that:

$p \rightarrow \mathbf{PnAntn}p$

Which is to say: If there is and was an omniscient being, then *everything has been anticipated.* And my contention, developed over the next few chapters, is that we can make do without any principle such as Retro-closure.

6
Betting on the Open Future

The overall aim of the book thus far has been to motivate and defend a conception of the open future on which future contingents systematically come out *false*. The current chapter, however, marks a transition in this book from (so to speak) the logical to the practical. On the one hand, the doctrine of the open future has always had its fair share of *logical* or *semantic* problems—roughly, problems associated with how the theory interacts with the principles of bivalence and Excluded Middle. In the previous four chapters, I have tried to address these problems by developing a theory on which the future is open (future contingents all fail to be true), but that retains both bivalence and LEM. However, these logical problems are not the *only* problems faced by the doctrine of the open future. Indeed, it seems to me that the chief philosophical objections to the open future are often not precisely *logical*, but instead roughly *practical*. For instance, how is the open futurist to make sense of a future-oriented practice like betting? Relatedly, how is the open futurist to make sense of *probabilities* of indeterministic future events, having insisted that all claims to the effect that any such particular event will happen fail to be true? (Of course, these problems are not strictly independent of one's answer to the logical problems.) It is these problems I begin to take up in the current chapter, and in the chapters to come.

I should begin with a word about my dialectical aims in this chapter. My aim is to defuse these objections to the doctrine of the open future—or, failing that, my aim is to show that these problems are no worse for my own version of that doctrine than they are for more familiar such versions. Because it seems to me that these problems are in the first instance *general* problems for any theory on which future contingents are *not true*, I will proceed mostly by employing 'not true' throughout, rather than 'false'.

6.1 Prior on Bets, Guesses, and Predictions

The first problem I wish to address concerns our practices of *betting*. The objection to be considered is not that the very fact that we have been betting is inconsistent with the claim that the future has been open. The objection is that, on the open futurist's theory, we are unable to make philosophical sense of that practice. That is, the objection, roughly, is that even if the open futurist's theory were true, in *explaining* the practice of betting, we would have to assume that it is

The Open Future: Why Future Contingents are All False. Patrick Todd, Oxford University Press. © Patrick Todd 2021.
DOI: 10.1093/oso/9780192897916.003.0007

false.[1] The following passage from A.N. Prior (himself an open futurist) provides a nice point of departure:

> [The] way of talking that I have just sketched [on which future contingents all come out false] shares with the three-valued way of talking [on which they are neither true nor false] one big disadvantage, namely that it is grossly at variance with the ways in which even non-determinists ordinarily appraise or assign truth-values to predictions, bets and guesses. Suppose at the beginning of a race I bet you that Phar Lap will win, and then he does win, and I come to claim my bet. You might then ask me, 'Why, do you think this victory was unpreventable [determined] when you made your bet?' I admit that I don't, so you say, 'Well then I'm not paying up then—when you said Phar Lap would win, what you said wasn't true—on the three-valued view, it was merely neuter: on this other view of yours, it was even false. So I'm sticking to the money.' And I must admit that if anyone treated a bet of mine like that I would feel aggrieved; that just isn't the way this game is played.
>
> ("It Was to Be" [1976], reprinted in Fischer and Todd 2015: 320)

Prior is right. This is, indeed, not how the game is played, and there is indeed something absurd about any such open-futurist "refusal to pay". But what is absurd here, I contend, is *not* open futurism (on grounds that it licenses such a refusal). What is absurd is precisely the idea that the open future view would license such a refusal in the first place. There is nothing in open futurism that is inconsistent with our ordinary practice of betting. Prior, I contend, was simply wrong to worry that there is.

First we need to state at least slightly more carefully what the problem is meant to be, as articulated here by Prior. The problem, as I understand it, is that since, on the open future view, the proposition "Phar Lap will win the race" was not true at the time of the given bet, it follows that anyone who had *bet* that Phar Lap would win the race would fail to win the bet—even from the perspective of a time at which Phar Lap has in fact won. Prior's hidden premise is perhaps obvious. Prior is hereby assuming that a *bet* that Phar Lap will win the race is in effect a bet *on the truth of the proposition* that Phar Lap will win the race; thus, since when one bets on Phar Lap's victory, one places a bet on the current truth of the corresponding proposition about Phar Lap, the open future view would license a refusal

[1] I mention this point because it is non-trivial move from the fact that we have to presuppose something in conflict with the open future when *explaining the practice* of betting to the claim that the open future view is *false*. In other words, an open futurist could simply insist that, when we are engaged in the practice of betting, we have to make certain "Ockhamist" assumptions—but that these assumptions do not in fact need to be *true* (only *thought to be* true) in order for us to have a successful betting practice. This may be so, but my aim in the following is to explain why, even in explaining philosophically the practice of betting, we needn't make any "Ockhamist" assumptions.

to pay—since, on that view, the proposition that one was betting was true was not, in fact, true. Hence the problem. The challenge is thus to give an interpretation of the practice of betting on which betting need not be construed as betting *on the current truth* of a claim about the future. And this is just what I aim now to provide.

6.2 Betting as a Normative Act

The basic outline of my reply to Prior's worry is simple; turning this outline into a rigorous theory is of course more difficult. The basic outline builds on ideas articulated by Hartshorne, and also Belnap and Green.[2] The basic idea is that, when it comes to betting on future events which have not yet transpired, *current truth* is irrelevant. Consider Prior's example of betting on Phar Lap to win the race. My suggestion is that when one bets on Phar Lap to win, one is not betting (or at least *need not* be betting) on the current truth of the proposition "Phar Lap will win". Rather, one is directly *bringing it about* that any future in which Phar Lap wins the race is *thereby* a future in which one wins the bet—and one can do this while fully acknowledging that there is no fact of the matter concerning whether Phar Lap will win. In other words:

> One *bets* that it will be that *p* iff one does something that brings it about that any *p* future is a future in which one is owed the [contextually specified] betting response.

And it would seem that one could rationally *bet* that it will be that *p* in *this* sense even if one believes that there is no truth, now, concerning whether a *p* future is the *actual* future. Thus, if I bet that it will be that *p*, and someone then points out that that there is no fact of the matter about whether any *p* future is *our* future (or that there is no such thing as *our future*), we can rightly say that this is irrelevant to the bet. Similarly, if anyone points out, after the fact, that on open futurism, it doesn't follow from the fact that *p* is true that a *p* future *was going to be* our future, we can similarly rightly say that this too is irrelevant. By betting, I bring it about that any *p* future is a future in which I am owed the betting response—regardless of whether, in that future, we can rightly say that that future was ours all along.

Look at it this way. Suppose we bet £5 on rain tomorrow. I bet on rain; you bet on no rain. But the matter is indeterministic. Nevertheless, *having now bet*, the causally possible futures are divided between rain futures in which I win, and no-rain futures in which you win. *This much is guaranteed.* There is no fact of the

[2] Hartshorne 1965: 54: "'Truth' is irrelevant to a wager"; Belnap and Green 1994: 383–384.

matter, now, concerning who will win—but, once more, this is irrelevant. The important point is that futures in which there is rain are futures in which I win; what is not relevant is the observation that, on the open future, just because it rains, it doesn't follow that it was *going to* rain. So suppose it rains. If you point out that back when we bet, a rain future was not yet *our* future (that is, it wasn't yet the case that there would be rain), I will reply that this doesn't matter: for the agreement was that any rain future is a future in which I win. We didn't agree that any rain future in which there had been going to be rain is a future in which I win. We *simply* agreed that any rain future is one in which I win. And we are, now, in a scenario with rain. So I win. It wasn't the case that I was *going to* win—but, to beat a dead horse, that simply doesn't matter.

6.3 Promising

But someone might already wish to press the following objection: what is it to bet "on rain" tomorrow, if not to bet *that it will rain* tomorrow? And what is it to bet *that it will rain* tomorrow, if not to bet on the proposition that it will rain tomorrow?

I am prepared to grant that a bet "on rain" tomorrow can be construed as a bet that it will rain tomorrow. What I deny is the second step, a move from the fact that I have bet that it will rain tomorrow to the claim that I have been on the truth of this *proposition*.

I recognize that these distinctions can seem both artificial and strained. In this respect, we might profitably consider other normative practices that are also sometimes future directed. Consider the case of *promising*. Suppose someone says, "Yes, OK, fine, I promise I'll pick you up from the airport tomorrow." Promising in this way is not best construed (or anyway needn't be construed) as saying something like, "Yes, I promise that it will be the case tomorrow that I pick you up from the airport"—which is an odd sort of promise to make, a promise that a claim about the future is true. The person is instead creating a new normative situation, one in which her failure to pick you up in a certain set of conditions for a certain set of reasons constitutes a moral violation. In other words, consider the oddness of the following:

A: OK, fine! I promise, I'll pick you up from the airport tomorrow.
B: That's a relief, thank you.
[A few hours pass, B abruptly decides to cancel her trip, informs A about her updated plans, with the result that the next day A never even attempts to pick up B from the airport:]
B: You know, yesterday you promised me that something was true that was false.

A: What? Really?

B: Remember, you promised me that it would be the case today that you pick me up from the airport. And yet, you aren't even at the airport.

A: When I promised that I would pick you up from the airport, I wasn't promising, then, that a certain proposition about the future was true!

B: Well, what else is it to promise that you will do something than to promise that the proposition that you will do it is true?

A: I don't know! God, what kind of philosopher are you?

The point is simple. One can promise that one will pick someone up from the airport without thereby promising that the proposition that it will be the case that one picks that person up from the airport is true. Similarly, one can bet that there will be rain tomorrow, without betting on the current truth of the proposition that it will be the case tomorrow that there is rain. It is only the contingent *idiom* with which we make promises that can make it seem, at a surface level, like A is promising B that something *is the case.* Similarly, it is, I suggest, only the contingent idiom—understandably so, given the Ockhamist presuppositions of ordinary discourse—with which we typically make bets (about the future) that can make it *seem* like, in betting, we are betting that something *is the case.* At a more fundamental level, however, when you bet that there will be rain, you are not betting that something is the case at all: again, you are instead engaging in a normative act that (as Belnap and Green say) has consequences for the future *however it unfolds*; you are creating a normative situation in which (*inter alia*) you are entitled to a certain response in certain conditions (and not in others). Again: a bet that there will be rain tomorrow is not really a bet (on the proposition) that *there will be rain tomorrow* at all. Indeed, as I aim to bring out shortly, we can sensibly bet *on rain* while explicitly *denying* that the claim that it will rain is true.

Of course, it is worthwhile to note that someone *can* indeed promise that a claim about the future is true—for instance, that it will rain tomorrow. (Perhaps someone wants to provide some assurance about the weather tomorrow.) But then there is, I contend, no problem for the open futurist arising from promises of this kind. If the matter is currently open, what this person promises is true is actually currently not true (or, as on my view, false). If you say, "It will rain tomorrow: I promise!", then if the matter is open, then you have promised that something is true that isn't. But this is no embarrassment for the doctrine of the open future; it is instead something more like a statement of the doctrine of the open future.

Similarly, we of course *can* bet, if we wish, on whether a certain claim about the future is now true. You can wager that a certain proposition about the future is currently the case; I can wager that that proposition about the future is not currently the case. But then notice precisely the plausible upshot of this bet. *This* bet indeed *does* implicate precisely all of the metaphysics our *normal* practice of betting completely bypasses. If we indeed do place a wager on the *current truth* of a claim about the future, then it indeed would be reasonable for an open futurist to refuse to grant the payoff, given a proof of past indeterminism; at any rate, whether this posture would be reasonable would then a *philosophical argument*. Even those of us not sympathetic to the doctrine of the open future nevertheless may be prepared to grant that there is at least a philosophical argument to the effect that future contingents fail to be true—but the point is that these debates would seem completely irrelevant to our ordinary practice of betting. We do not need to settle these debates in order to determine whether someone who had bet on Phar Lap is owed the payment if Phar Lap wins; however, if the bet on Phar Lap must be construed as a bet on the truth of the relevant proposition, these debates would indeed be implicated. But they aren't. So the bet was not a bet on the current truth of a proposition about the future.[3]

6.4 Open-Futurist Agreements

There is a further way to observe that betting on the future needn't be construed as betting on the truth of claims about the future. And that is that such bets would seem to make sense, even in the face of explicit avowals that the future is *open*. Consider what we might call the *Open-Futurist Agreement*:

A: It is not now true that there will be rain tomorrow, and not now true that there will be no rain tomorrow. It is open. But let's agree: if it does rain tomorrow, you owe me £5, and if it does not rain tomorrow, I owe you £5. Agreed?
B: Agreed.

As I see it, there is nothing in the least bit incoherent or problematic about this agreement; indeed, it seems to be a paradigm of intelligibility. But now notice the

[3] Witness, on this point, the difference between two different games one might encounter in Las Vegas. The first is roulette. The second is what we might call "Now-true roulette"—a game that might be advertised (a bit ham-fistedly) as follows: "Place your chips next to the following options to bet on the current truth of the proposition that on the next spin the ball shall land in the slot associated with that option". My sense in that *this* game may rightly have fewer takers than the former. It is certainly not obviously equivalent to the former; at least, the question of their equivalence is a confusing question. (NB: it would seem that these games are not equivalent, even without appeal to open-futurist metaphysics; note that the casino can't win in roulette by spinning the ball, but then ensuring that the ball is destroyed before it ever lands in a slot, but a casino arguably *could* win the latter game in this fashion, even by the Ockhamist's lights. These points are related to Dummett's distinction between a conditional bet and betting on a conditional, but I set these points aside.) Of course, non-open futurists might contend that the only reason they wouldn't want to play this latter game is simply that, as they see it, confused open futurists might be inclined to cheat them out of their winnings by invoking their bizarre thesis that the relevant propositions weren't in fact true at the time of the given bets. Be that as it may, the point remains that insofar as we do grant a difference in these two games (the open futurists could raise such hackles concerning the second, but not the first), then this is enough to respond to the worry as envisaged by Prior. The point is that if we're playing roulette, past truth is neither here nor there; if we're playing Now-true roulette, past truth is exactly what is at issue.

following. If one insists that a bet on rain tomorrow must be construed as a bet on the truth of the proposition that there will be rain tomorrow, the Open-Futurist Agreement would seem to amount to the following:

A2: It is not now true that there will be rain tomorrow, and not now true that there will be no rain tomorrow. But let's agree: if (I am mistaken and) it *is* true that there will be rain tomorrow, then you owe me £5, and if (I am mistaken and) it is true there will be no rain tomorrow, then I owe you £5. Agreed?
B2: I'm confused.

But the point is clear. In the original agreement, the participants were not betting on the proposition that they are currently mistaken (and the future isn't really open after all). Of course, it is not impossible to bet on that proposition, but the point is that that is not what they were doing. The agreement is not that though it is not true that it will rain, if it *is* true that it will rain, then someone is owed some money. The agreement is not about current truth; it is about obligations to pay in various scenarios that might unfold.

My own position once more is the following. The reason why we don't hear the Open-Futurist Agreement as embodying anything like a contradiction (or otherwise "Moore-paradoxical") is that we implicitly interpret the "But if it does . . . ", not as concerned with anything like *truth* ("But if something is true . . . "), but as a way of specifying the scenarios under which the bet is won or lost. In other words, we have something like the following:

A: It is not now true that there will be rain tomorrow, and not now true that there will be no rain tomorrow. It is open. But let's agree. Futures with rain tomorrow shall be futures in which you owe me £5, and futures with no rain tomorrow shall be futures in which I owe you £5. Agreed?
B: Agreed.

And now there is nothing even *prima facie* contradictory about the relevant agreement. And if there isn't, Prior's worry (as articulated above) is misplaced. There is a perfectly intelligible account of betting on which betting is not to be construed as betting that a certain proposition is currently the case—and if this is so, the argument against open futurism entertained by Prior fails.

6.5 Ambiguities

Let me pause to note one potential source of confusion arising from Prior's challenge as presented above. Recall:

> [The] way of talking that I have just sketched [on which future contingents all come out false] shares with the three-valued way of talking [on which they are neither true nor false] one big disadvantage, namely that it is grossly at variance with the ways in which even non-determinists ordinarily appraise or assign truth-values to predictions, bets and guesses.

Note that Prior (perhaps slightly waffling on the precise nature of the problem) says "appraise *or* assign truth-values". However, we might rightly insist that these two acts can come substantially apart. One can, I contend, rightly *appraise* a bet as having won, *without* assigning the truth-value "true" to the antecedent claim that it was *going to win*. Similarly, one can appraise a guess or a prediction as having been *vindicated* or *impugned* (as in Belnap and Green), without assigning the truth-value "true" to the antecedent claim that it was *going to be* vindicated or impugned. As I see it, there is a sense in which "bets, guesses, and predictions" can be assigned truth-values, and a sense in which they might merely be *appraised* in the manner indicated above.

The terminology of "bets, guesses, and predictions," however, is often ambiguous between these different senses. Consider, for instance, the following statements, all of which can seem (given the right context) perfectly appropriate:

(a) I'm betting on red, but I bet I'll lose this bet.
(b) I bet on Phar Lap, but I certainly wasn't betting that he'd win.
(c) Needing to avoid suspicion, betting that Phar Lap would win, he bet that Phar Lap would lose.

In these cases, the underlined usages refer to the given agent's mental state, and mean (something like) "think/ing that"—and *here* the question of truth or falsity (or neither) can appropriately arise. (What the person is thinking may be true, false, or neither.) The non-underlined usages, however, refer to one's bet *in the relevant betting practice*. Something similar may occur in the case of guessing:

(d) I'm guessing red, but I'm guessing it won't be red.
(e) Needing to avoid suspicion, Anders guessed red, guessing that it wouldn't be red.

Finally, we can observe something similar regarding prediction:

(f) Doug's prediction for the newspaper was that Phar Lap would win, since he was predicting that Phar Lap would lose.

The point: betting, guessing, and predicting in *these* (non-underlined) ways are *normative acts*. When betting, guessing, and predicting in this way, one is not attempting to match one's belief to a pre-existing fact in the world, but rather *creating* a new circumstance *in* the world. However, the underlined usages are different—and here Prior *could* be alluding to a distinct challenge for the open futurist, a challenge directly to the effect that the open futurist cannot assign the truth-value "true", for instance, to Doug's *prediction* (read: belief) that Phar Lap would lose, despite the fact (say) that Phar Lap has in fact lost. But if there is a philosophical problem arising for the open futurist from *this* observation, it is simply the problem, addressed in Chapter 7, that open futurists must deny what many regard as an intuitive principle of tense-logic, viz., $p \rightarrow \mathrm{PnF}np$. And thus I postpone discussion of this problem until Chapter 7.

6.6 Omniscience

At this stage, I wish to put to work the points developed in the previous chapter concerning the analytic connection between *truth* and the beliefs of an omniscient being. It is here, I think, that it becomes most apparent that (as Hartshorne says) "truth is irrelevant to a wager". (These connections will also prove crucial in responding to the further problems to come.) Simply put: insofar as we are prepared to admit that the current beliefs of an omniscient being are irrelevant to a wager—and I think that we are—we should be prepared to admit that truth is similarly irrelevant. Consider the following:

A: Let's bet £5 on rain tomorrow. If there's rain, you owe me £5, and if not, I owe you £5. Deal?
B: Deal.
[A day passes, and there is rain]
A: You owe me £5!
B: Not so.
A: And why is that?
B: Well, yesterday there was an omniscient being—God—and yesterday that being didn't anticipate rain today. Isn't that right, God?
GOD: Yes, that's right. With respect to rain, yesterday I had no anticipation concerning what would happen today; I didn't anticipate rain, although of course I didn't anticipate the absence of rain.
B: And, God, yesterday you were omniscient, weren't you?
GOD: Of course.
B: So, yesterday, it wasn't true that there would be rain today.
GOD: Yes, that's right.
B: So I'm not paying up; what A bet was the case, as we have seen, wasn't the case.

B's posture in this dialogue is, of course, absurd—but the trick is accurately to diagnose from whence the absurdity comes. One option is to maintain that B's posture is absurd on grounds that B's—and God's—*philosophical view of truth* is absurd; they are both mistaken to maintain that God could have been omniscient yesterday, despite not anticipating today's rain. B is not wrong about the nature and content of A's bet; B is simply wrong to allege that the proposition A was betting was true wasn't in fact true. But even if such an explanation were available, such an explanation, in my view, is simply not needed. A should respond, in the first instance, not by challenging God's claim to having been omniscient (although I grant that that may be done), but by challenging the relevance of the supposition that God *was* omniscient to her winning the bet. On my view, that is, A should respond as follows:

> A: When yesterday I was betting on rain today, I plainly wasn't betting something like this: "I hereby bet that any omniscient being who might exist currently anticipates rain tomorrow." Insofar as I wasn't, I was likewise not betting something like this: "I hereby bet that it is currently true that there will be rain tomorrow." So the fact that yesterday God didn't anticipate rain today is irrelevant, and so similarly is the fact that yesterday it wasn't true that there would be rain today.

A's posture is reasonable. B's conception of truth (and God's claim that yesterday he was omniscient) are also reasonable. But even if that is so, B's refusal to pay isn't.

The case of an omniscient being further helps to dispel the sense that, on the open future, there is nothing to place a bet *on*—after all, *ex hypothesi*, it is not true that it will rain, and not true that it will fail to rain. And if this is so, and we know it, where is the room for betting? But consider:

> A: Let's bet £5 on rain tomorrow. If there's rain, you owe me £5, and if not, I owe you £5. Deal?
>
> B: Well, unfortunately, there's nothing really to bet on here. For there is an omniscient being—God—and right now, God does not anticipate rain tomorrow, and God does not anticipate an *absence* of rain tomorrow. Isn't that right, God?
>
> GOD: That's right.
>
> B: So unfortunately, any betting here makes no sense. If you were going to bet on rain tomorrow, you would have to be betting that God currently anticipates rain tomorrow. And if I were going to bet on no rain tomorrow, I would have to be betting that God currently anticipates no rain tomorrow. But, as we have just seen, God has no anticipation either way at this stage: and so it would be senseless for us to place bets on what God's anticipations are on this matter; God has no such anticipations.

A: Well, OK—that may be senseless, given what God just said, but why do we have to place bets on God's anticipations? Why can't we just place bets on rain tomorrow?

B: But look, placing a bet "on rain" tomorrow is equivalent to placing a bet on the claim that it will rain tomorrow, which is in turn equivalent to placing a bet on the claim that any omniscient being anticipates rain tomorrow. So, again, there is no space to simply bet "on rain" without betting that any omniscient being anticipates rain.

A: Let me try again. We don't have to place bets on God's current anticipations regarding rain tomorrow. We can simply, by saying some relevant words, *agree* that any scenario with rain tomorrow shall therefore be a scenario in which you owe me £5, and any scenario with no rain tomorrow shall therefore be a scenario in which I owe you £5–regardless of whether, in the envisaged scenario, God had previously anticipated rain or instead no rain. Shall we agree?

B: OK. We shall.

And all is well. A and B have bet on rain tomorrow, but have no need for facts about whether it will rain tomorrow.

6.7 Transition to the Credence Problem

Thus far we have been concerned simply to reply to the problem for the open futurist articulated by Prior. But now consider the following associated problem articulated by MacFarlane—a problem, I think, that raises substantially more serious difficulties:

> [Consider what] we might call the *Credence problem*. The [open futurist] predicts, plausibly, that an agent who thinks both a sea battle and peace are possible should not believe either that there will be a sea battle or that there will not be a sea battle. (Both propositions are untrue as used and assessed from the agent's context.) But in addition to asking about full belief, we can ask about partial belief. What credence (subjective probability) should the agent have in the proposition that there will be a sea battle?
>
> Here competing considerations seem to point in different directions. On the one hand, the agent knows that the proposition that there will be a sea battle is not true. Normally we give a very low credence to things we are certain are untrue. That suggests that the agent should have a very low (perhaps 0) credence in both the proposition that there will be a sea battle and [the proposition that there will not be a sea-battle].
>
> On the other hand, when an agent has a credence of 0 that *p*, we generally take it to be irrational for her to accept a bet on *p at any odds*. (The expected utility of

> the bet is the value of the bet times the credence in p minus the cost of the bet; so if the credence in p is 0, then the expected utility of the bet cannot be greater than 0.) So, if the considerations in the previous paragraph are correct, an agent who thinks that both Heads and Tails are objectively possible outcomes of a coin flip should not accept a bet at any odds on the outcome. And that is surely wrong. Surely it would be rational for the agent to pay one dollar for a chance to win a hundred dollars if a fair coin lands Heads, and irrational to decline such a bet.
>
> So we have a dilemma. Either we preserve the connections between degree of belief and truth—the idea that, when one is certain a proposition is not true, one believes it to degree 0—or we preserve the connections between degree of belief and rational action (for example, in accepting bets). We cannot have both, but both seem essential.
>
> One might think that the dilemma can be resolved by observing that the rationality of betting depends on the chance one will get the payoff. This, in turn, depends not on whether the proposition that the coin will land Heads is true as assessed from the present moment, but whether it will be assessed as true after the coin lands, when the payoff will be determined. So, the fact that the proposition is neither true nor false as assessed from the present moment is irrelevant to the rationality of the bet. But this doesn't really help, since the proposition that *is* relevant to [the rationality of] the bet—namely, that the proposition that the coin will land Heads will be assessed as true after it lands—is itself a future contingent. (2014: 233–234)

MacFarlane's own resolution of this problem depends on the details of his own particular version of relativism (assessed in the next chapter). But I do not think we need to have recourse to relativism to solve this problem. My claim is parallel the claim I made above. The proposition that MacFarlane says *is* relevant to the rationality of the bet is not relevant to the rationality of the bet. In short, if, as we have seen above, the truth of future contingents is not presupposed by one's *winning* a bet, it is likewise not presupposed for the rationality of *placing* a bet. Let it be granted: the proposition "*The coin will land Heads* will be assessed as true after it lands" is itself a future contingent. And thus let it be granted that, on open-futurist grounds, one's credence in that proposition—more simply, the proposition that the coin will land Heads—ought to be 0. It does not follow, I contend, that a bet on Heads is, for this reason, irrational.

Let me first put my cards on the table. It is when we confront what to say about probability that the open futurist must be at his or her most revisionary; as we'll see shortly, in order to make sense of there being non-zero probabilities of future events, the open futurist seemingly has to say some strange things. Indeed, some who are otherwise attracted to the views defended in this book may despair at the results to come. But there are at least two reasons why despair is the wrong attitude to adopt in the face of these results. The first is that probability is one of the most intractable notions in all of philosophical theorizing, and it is of some comfort that no one knows how to make sense of it, let alone how to make sense of it in conjunction with a revisionary metaphysical hypothesis to the effect that future contingents systematically fail to be true. The second is that whereas the open futurist must endorse some claims in this domain that will strike many as revisionary, there is, I contend, no reason to regard these claims as incoherent. They are indeed coherent, and they are motivated by a reasonable metaphysical picture of the world; that reasonable metaphysical picture, like many others, does force us to say what we do not find natural to say. But it is wrongheaded to reject this metaphysical picture *on these linguistic grounds*. It is a mistake to postulate a realm of future-facts solely to avoid having to make a few unfamiliar—but nevertheless coherent—semantic distinctions.

6.8 The First Problem: Zero Credence

The first problem is that if my credence in the claim that there will be rain tomorrow is 0, then it would be irrational to accept a bet on rain tomorrow at any odds. I don't think this follows, but it is a complicated affair explaining why it doesn't. More particularly, I think that this follows only on *Ockhamist* assumptions about the nature of the facts about the future. For, on those assumptions, if my credence in the claim that there will be rain is 0, then I will therefore be certain that that claim is false—and if I am certain that that claim is false, I will therefore be certain that it will be tomorrow that there is no rain. (On Ockhamist assumptions, ~Fn*p* plainly entails Fn~*p*.) And if I am certain that it will be tomorrow that there is no rain, then of course it would be irrational to place a bet on rain tomorrow. This much is clear. But matters become substantially more complex if we can maintain, as I have done above, that ~Fn*p* does not entail Fn~*p*. For then, perhaps, my credence in Fn*p* could rightly be 0, but nevertheless my credence in Fn~*p* could rightly fail to be 1; indeed, my credence in that claim could *also* be 0.

That one could rationally place a bet on rain while having credence 0 in the claim that it will rain is certainly strange, but consideration of the beliefs of an omniscient being, once more, can make it seem less so:

A: God, I'm considering a bet on rain tomorrow; I just have to pay £1 for a chance to win £1000 if there's rain. But, since you're omniscient, I'm curious: what are the chances that if I bet on rain, you'll thereby anticipate my winning that bet?

GOD: Zero.

A: Oh. That's disappointing. But maybe I should also ask, since I'm in a philosophical mood: what are the chances that if I bet on rain, you'll thereby anticipate my *losing* that bet?

GOD: Also zero.

A: I see! So, if I bet on rain, you'll have no anticipation either way regarding my winning or my losing.

GOD: That's right. I have no anticipation regarding rain or no rain tomorrow, and your bet certainly wouldn't have an effect on this matter—so if you bet on rain, I'd have no anticipation regarding your winning or your losing.

A: And you're omniscient?

GOD: Of course.

A: Got it. I think I'll take my chances.

[A bets on rain]

A presumably now flatly accepts the claim that God does not now anticipate her winning the bet tomorrow. Since A also accepts that God is *omniscient*, A should now be prepared to grant that it isn't *true* that she will win her bet tomorrow. That sounds bad for A. And yet: A also accepts that God does not now anticipate her *losing* the bet tomorrow, and therefore also accepts that it isn't true that she will *lose* her bet tomorrow. Thus far, in none of this is it the case that A is being irrational. She rationally took the bet, even while fully accepting that if she takes the bet, it won't be true that she'll win. But this is because she also accepted—what only an open futurist could then accept—that if she took the bet, it wouldn't be true that she would *lose*.

Let me cut to the chase. On the open-futurist picture I defend, if the future is open with respect to rain tomorrow, then if you bet on rain tomorrow, there is no chance at all that the proposition that you will win the bet will be true. However, crucially, it does not follow that it would therefore be irrational—harkening back to the model of betting defended above—to *stake a claim* to the rain futures. I think there is no fact of the matter whether it will rain—and yet, here is my chance, not to wager that the fact of the matter is that it will rain, but instead to *claim* the rain futures as my own, *regardless* of whether it is true that any such future is already "our future". Why should the fact that there is currently no fact of the matter concerning rain tomorrow imply that it would be irrational to stake a claim to the rain futures—especially if I only have to pay £1 to do so, and they thereby become futures in which I'm owed £1000?

But here we come to the fundamental issue. For simplicity, reconsider the case of a simple wager of £5 on rain tomorrow. What should help to determine whether I should make this bet? Not, I have suggested, my beliefs about the chance that a rain future is our "actual" future. But then what? More particularly, I claim, there is no reason, in this scenario, to regard it as irrational for me to stake a claim to the rain futures, *if I think that current reality is tending towards the realization of one of those futures, even if there is no fact of the matter concerning how these tendencies will be resolved*. What matters to the rationality of this bet, in short, is not the likelihood of truth of any given future contingent; what matters is instead the strength of the world's current *tendencies*. And it is fundamental to the open futurist's picture of reality that the world could be strongly *tending* in a certain direction, *without* this implying anything about a likelihood of a current *fact* about the resolution of those tendencies. There are the tendency facts, but no *further* facts about the *resolution* of these tendencies. Since there *are no* such facts, claims purporting to *report* such facts—claims, for example, to the effect that there will be rain tomorrow—are claims in which I will have accordingly have credence 0.

Clearly, what the open futurist needs here (*inter alia*) is some idea about the fundamental ontology of the world on which reality (or its components, or ...) has indeterministic *tendencies*. But this is, I suggest, already part and parcel with the open futurist's basic outlook: the open futurist contends that, at the fundamental level, what exists is the present (and perhaps also the past) and the laws; what currently exists, as governed by the laws, may have *tendencies* or *dispositions*, but these tendencies or dispositions are not always deterministic. The current chapter is certainly not the place to develop or defend this more basic ontological picture of reality and law; as far as I can see, this picture could be developed in at least several competing ways, all of which are nevertheless consonant with the basic picture as stated above. What is important here is that the open futurist can defend a picture on which reality can *tend* towards certain outcomes with a certain degree of causal strength, and that this degree of strength can be represented as being between 0 (no tendency at all; that is, the event is determined to fail to happen) and 1 (the event is determined to happen). Herewith I assume that the open futurist is entitled to just such a picture.[4]

Notice how starkly the resulting view must contrast with a more standard, common-sensical "Ockhamist" picture. On Ockhamism, there are of course the underlying causal tendency facts. However, for the Ockhamist, in addition to these tendency facts, there are *also* the facts (standardly unknowable by human beings) about how these tendencies eventually will be resolved. Accordingly, a rational (non-God-like) agent will use her estimate of what the underlying tendency facts are to form an estimate of what the *facts of resolution* will be. If the agent thinks that the underlying tendency facts are that the world is tending towards rain with strength .6, that agent will naturally think that there is a .6 chance that the *further fact* is that there will be rain, and will therefore naturally assign a .6 credence to the claim that there will be rain (that is, be .6 confident [however this is to be understood] that that claim represents a truth). But for the *open futurist*, if we think that the underlying tendency facts are that the world is tending towards rain with (merely) strength .6, we will think that there is no *further fact* that there will

[4] For recent discussion of one such picture of tendencies, see Anjum and Mumford 2018; though note their complications about the relationship between tendencies thus conceived and probabilities (57–58). Note: I am certainly not saying that the relevant picture of indeterministic tendencies is easy to provide; I am simply saying that, however hard or easy this may be, this is what is required.

be rain, and will therefore consider that there is a 0 percent chance that the further fact is that there will be rain, and will therefore naturally (or, to be fair, not very naturally) assign a credence of 0 to the claim that there will be rain. The result is that, somehow, we think that rain tomorrow is objectively likely, but we have *no degree of confidence at all* in the *claim* that there will be rain. This is certainly strange. But there it is.

The intended result, and the result to be defended, is thus the following. When the world is tending toward rain tomorrow with strength .6 and towards non-rain with strength .4, and there is no fact of the matter, now, concerning whether there will be rain or instead no rain, then:

The probability that it is true that it will rain = 0
The probability that it is true that it will fail to rain = 0

And on my own view, we can add:

The probability that it is false that it will rain = 1
The probability that it is false that it will fail to rain = 1

But we also must find a way of saying:

The probability of rain tomorrow = .6
The probability of no rain tomorrow = .4

And we must do this without contradicting ourselves—that is, without saying that there is a .6 chance that something that is untrue or false is true. In other words, we must distinguish between the probability *of the claim* (the future contingent) that it will rain tomorrow from the probability of rain tomorrow. We must distinguish between the strength of the world's tendency to produce a certain outcome tomorrow—viz., *rain*—and the likelihood of the claim that there will be rain. The current causal tendencies of the world can make rain tomorrow likely, but *not* make likely the truth of the proposition that there will be rain tomorrow. And we must insist that when you are *betting* on rain tomorrow (or, for that matter, deciding whether to take an umbrella for tomorrow), what matters is not the probability of the *claim* that it will rain tomorrow, but instead the probability of rain tomorrow—and whereas the former is low, the latter is high, for whereas the former is tied to a likelihood of *current truth*, the latter is tied to the strength of the world's *tendencies*.[5]

[5] In conversation, several interlocutors have suggested that, on the view I defend, one should tie the probability of rain tomorrow to the *proportion of available futures* in which there is rain tomorrow; to say that rain is 60% likely, so the suggestion goes, is merely to say that there is rain in 60% of the

Let me further note the following. Someone proceeding on these assumptions, I contend, will be observationally equivalent to someone proceeding on Ockhamist assumptions; that is, the two agents will make all and only the same bets, but they will simply diverge in the *explanations* of the justification of their bets. When betting on rain (viz., claiming the rain futures), the open futurist will explain that, though it isn't true that it will rain, it also isn't true that it will fail to rain—but at any rate, the world is certainly *tending toward* rain with a certain strength, e.g., strength .6. When making the same bet, the Ockhamist will explain that, though it isn't causally determined that there be rain, it isn't causally determined that there fail to be rain—but at any rate, the world is certainly *tending toward* rain with a certain strength (.6), and therefore it is (to that degree, i.e., .6) likely that the fact is that it will rain. The open futurist simply refrains—on ground of ontological parsimony—from making this latter addition. The latter addition may be independently plausible, but *it is not required for the rationality of betting.*[6]

6.9 Moore-Paradoxes?

Recall what we above called the *Open-Futurist Agreement:*

A: It is not now true that there will be rain tomorrow, and not now true that there will be no rain tomorrow. It is open. But let's agree: if it does rain tomorrow, you owe me £5, and if it does not rain tomorrow, I owe you £5. Agreed?

B: Agreed.

available futures. (For a suggestion of exactly this kind, see Longenecker 2020: 1498.) However, I am dubious that this suggestion can work, precisely because the following scenario sounds, to me, perfectly coherent: there are two possible ways things could be tomorrow, but one is more likely to obtain than the other. On the suggested view, however, this would be incoherent: if there are two and only two possible ways things could be tomorrow, then the relevant probabilities would have to be 50/50. For example: it seems perfectly possible that there should be, now, only two remaining candidate ways things could be tomorrow: a scenario in which a given coin lands heads, and one in which it lands tails; everything else is pre-determined. But it could nevertheless be that the coin is *weighted* (causally speaking) towards heads. In general, the fact that the causal tendencies (now) favor rain tomorrow does not imply that there are "more" causally possible scenarios in which it rains tomorrow than causally possible scenarios in which doesn't.

[6] Let me make a note of one further strategy the open futurist might employ to lessen the shock of the claim that one's credence in the claim that there will be rain tomorrow ought to be 0. As before, we consider the connection between truth and the beliefs of an omniscient being. Now ask yourself: given the openness of the future, what is your credence in the claim that an omniscient being (supposing one exists) anticipates rain tomorrow? If we know that the future is open, here we may comfortably say '0'. (And so similarly for the claim that an omniscient being anticipates an *absence* of rain tomorrow.) But when we recognize that p is logically equivalent to q, plausibly our credence in p should match our credence in q. Thus: since *an omniscient being anticipates rain tomorrow* is logically equivalent to *there will be rain tomorrow*, our credence in the latter should likewise be 0. But we shouldn't then confuse our credence in this claim about rain tomorrow for the objective probability of rain tomorrow.

And my central contention was that this agreement is in no way Moore-paradoxical. But now consider a parallel statement, not concerning a *bet*, but simply about *chance* directly:

(1) It isn't true that it will rain, and nor it is true that it will fail to rain. It is open. But still, I'm putting the chances of rain at .6.

Now, *I* at least can make sense—or claim to be able to make sense—of this statement. Indeed, as I see it, this statement is easily intelligible; this person is claiming that whereas there is now no fact of the matter concerning whether it will rain tomorrow (nothing "there" for an omniscient person to know), nevertheless the objective probability (based on some imagined set of causal tendencies) of rain is .6. But contrast *that* statement with the following:

(2) It isn't true that it will rain, and nor is it true that it will fail to rain. But still, I'm putting the chances that it is true that it will rain at .6.

And here we have something that is obviously Moore-paradoxical. But if (1) was *not* Moore-paradoxical, then the "chances of rain" in the latter sentence of (1) must not be equivalent to the chance that "There will be rain" represents a truth. And this is the key claim we need in order to make sense of our future-oriented actions and practices in an open-futurist setting.

Consider further the following:

(3) It isn't true that it will rain, and nor is it true that it will fail to rain. It is open. But still, I think it will *probably* rain.

Again, this seems perfectly felicitous—and its felicity is strong reason to give some interpretation of "It will *probably* rain" that doesn't commit this speaker to the probability that she is mistaken.

6.10 The Second Problem: The Linguistic Data

But now we come to a difficult objection (indeed, the cause for despair alluded to above). The objection is that, in the scenario in question, I in particular am now committed to following unlovely claims:

(4) It is false that it will rain, but it is probable that it will rain.
(5) It is false that it will rain, but it will probably rain.

And so I am. But my contention is that the reason these claims *sound* unacceptable is because, for those who haven't read the last five chapters of this book, they sound like saying that though it won't rain, it is probable that it will. And of course I am *not* committed to the truth of that proposition—distinguishing, as I do, between its being false that it will rain and its being true that it will fail to rain.

I am, however, getting ahead of myself. Let me make one brief note before moving on. The argument against my view now under consideration is, once more, an argument from a judgment about linguistic "felicity". But one might, in general, have doubts about the strength of this sort of strategy to settle the sort of philosophical disputes about openness at issue in this book. Note, for instance, that "My children are at home" certainly seems to mean "All of my children are at home". Thus, "Most of my children are at home, but my children aren't at home" should sound fine—and yet it sounds terrible.[7] Likewise, I am prepared to admit that "It is false that it will rain, but it will probably rain" sounds terrible. But it is one thing to note that this claim sounds terrible; it is another to show that it couldn't possibly be true.

But I'll stop making excuses. My contention is that when *will* combines with "likely" or "probably", as in "It will probably rain tomorrow", the resulting sentence does not have what may *seem* initially like its simple logical form, viz.,

Probably: Fn[Rain]

Thus, (4) and (5) cannot be represented as anything like

$\sim \mathrm{Fn}p \wedge \mathrm{Probably{:}Fn}p$.

Rather, the logical form of "It will probably rain tomorrow" is instead ... well, something else. To a small primer on that something else I now turn.

6.11 The Will/Would Connection Once More

Once more, as it happens, the best way to defend my own view of the open future is to investigate what has *already* been said about counterfactuals (especially in the context of denials of Conditional Excluded Middle), and import the solution that has been offered in *that* domain to the current domain. (See Chapter 4.) And, unsurprisingly, when we look to the literature on counterfactuals, we find exactly the argument—an argument from credence—against the denial of CEM as has just been considered against my own denial of WEM.[8]

[7] Thanks to Wolfgang Schwarz for this example. As Schwarz notes [personal correspondence], the example may be more than a mere example; the underlying issues may very well be the same in these two cases—but I set this complex issue aside.

[8] For an excellent statement of this argument, see Mandelkern 2018; my presentation of the problem to come is partially based on Mandelkern's.

I begin with the following simple observations. Suppose we asked Williamson and Lewis the following (deceptively simple) question. What are the chances that had you flipped a perfectly fair coin yesterday, it would have landed heads? The right answer is presumably something like ".5" or "50 percent". The wrong answer, presumably, is "0". Surely there is a *chance* of a fair coin landing heads. But giving this answer, if you deny CEM, is a surprisingly difficult affair. After all, on the view in question, it is *false* that if you had flipped a fair coin yesterday, it would have landed heads. How then can there be a chance that it would have landed heads? Wouldn't this have to amount to a chance that something that this view says is false is true?

Perhaps we can best bring out the problem as follows. Consider:

> If I had flipped the indeterministic coin, it would have landed heads.
> If I had flipped the indeterministic coin, it would have landed tails.

Now, for the denier of CEM, both of these claims are *false*. But now consider, not a *perfectly* fair indeterministic coin, but instead a *slightly weighted* indeterministic coin; in particular, an indeterministic coin slightly biased towards heads. And consider:

(i) If I had flipped the slightly heads-biased indeterministic coin, it would have landed heads.
(ii) If I had flipped the slightly heads-biased indeterministic coin, it would have landed tails.

Now, it is crucial to observe that, for the denier of CEM, both of these claims are *also* false. But now the point. Presumably the denier of CEM will want—indeed, must have—some way of recovering the thought that a slightly heads-biased coin is more likely to have landed heads, had you flipped it, than a perfectly fair coin. But they have also just maintained that it is *false* that such a coin would have landed heads if it had been flipped. Thus, the denier of CEM certainly cannot attempt to capture this thought by saying that the former claim (i) is more likely to be true than the latter (ii); they have foregone that possibility by saying that both are false. And now you can see where we're headed. For it would seem that this combination of views is now committed to the claim:

> It is false that the slightly heads-biased indeterministic coin would have landed heads, but it would probably have landed heads.

And that does not—like (5) considered above—strike us as a readily intelligible thing to say. So what does the denier of CEM have to say for herself?

More generally, observe that, for a denier of CEM, we will have to say something like the following:

> The probability that the coin would have landed heads if it had been flipped *does not equal* the probability of the conditional "If the coin had been flipped, it would have landed heads."[9]

After all, the denier of CEM presumably wants to say that the former could be .5 or .6, whereas the latter is 0. And now my claim is simple. If we can make sense of *that* (and I hope we can) then we can make corresponding sense of the claim that the probability of rain tomorrow does not equal the probability of the claim "There will be rain tomorrow". Indeed, the former could be .6, whereas the latter could be 0.

My general point here is that if one denies *on these semantic grounds* that the probability of rain tomorrow can be .6, but the probability of the claim "There will be rain tomorrow" can be 0, then one must prepared to make a similar semantic argument in the case of counterfactuals; if the probability of the claim "If the coin had been flipped, it would have landed heads" is 0, then this thereby implies that there can be no objective chance that the coin would have landed heads if flipped. And if one thinks that it is a decisive objection to the current theory that it is committed to the possible truth of "It is false that there will be rain tomorrow, but it will probably rain tomorrow", then you must likewise be prepared to make the parallel semantic argument for CEM, from the observation that such a denial is committed to the possible truth of, "It is false that the coin would have landed heads if it had been flipped, but the coin would probably have landed heads had it been flipped." Once more: these semantic arguments *stand or fall together.*

6.12 Probability in Fiction

Let me end this chapter with a comparison that has, to my knowledge, gone unnoticed in the vast literature both about the open future, and about the probability of conditionals and conditional probability. Note that the problem for the denier of CEM seems to arise from the following tension: on the one hand, we want to say that there is no fact of the matter about what would have happened had we flipped a fair, indeterministic coin; we thus say that it is false that the coin would have landed heads, but also false that the coin would have landed tails. On

[9] Such a result is of course reminiscent of Lewis' much-discussed "triviality results" for indicative conditionals and conditional probability—a topic to which I shall not, in this chapter, even attempt to do minimal justice. For more on this subject, see Bennett 2003: ch. 5; and for more on this subject as it relates to counterfactuals in particular, see Williams 2012, and Moss 2013.

the other hand, we want to say, of course, that there is a good chance—indeed, a 50 percent chance—that such a coin would have landed heads. But we want to say this without saying that there is a 50 percent chance that something false is true; indeed, there is something that has a certain positive probability, but this is *not* to be identified with the truth of the relevant counterfactual. Similarly, the open futurist (on my account) wants to recover the idea that certain events have a positive, non-zero chance of happening, *without* saying that this corresponds to a good chance that something false is true. Something has some positive probability, but *not* the relevant future contingent.

It would thus seem to help my case if we could find some *third* domain where *everyone* might agree that the relevant "probabilities" have this kind of structure—and that is what I now aim to provide. Here we turn, then, to a comparison with probabilities and *fictions*. Here we can harken back to Chapter 1, and recall what is, for me, one of the controlling motivations for the doctrine of the open future. And that is that, on presentism and indeterminism, the facts about the future stand to the present and the laws as the facts about fictions stand to the *fiction-determining facts* (whatever they are). If we wanted to press this analogy in a certain direction, we might say that *we ourselves*—by acting in accordance with the laws—*write the story of the future*, but that the story of the future is not yet written; indeed, just as we might wish, it is being written *as we write it*. Leaving us and our own agency out of the picture, however, we can simply say, once more, that the present and the laws stand to facts about the future as the fiction-determining facts—whatever they are—stand to facts about fictions.[10]

And now a comparison emerges. For recall what we wished to say in Chapter 1. If we are reading the Harry Potter books, and in a certain episode Harry is having breakfast with some orange juice, someone might wish to ask: what was the precise temperature of that orange juice? And here all of us—or anyway nearly all of us—are inclined to say that there is no fact of the matter. There is no fact of the matter about the temperature of Harry's orange juice in the episode at issue; nothing "there" to know. Similarly, there is no fact of the matter concerning precisely how many soldiers took part in the Battle of Helm's Deep in the *Lord of the Rings* saga. We can say definitively that it wasn't 100,000,000—for that would contradict the

[10] A similar comparison has been made regarding counterfactuals; cf. Hare 2011: 193:

> After you have declined to spin the wheel, it will sound right to say "I don't know what would have happened if I had spun it." Once you have said that, you may be tempted to infer that there is something to be known that you don't know: either you would have gotten a red, and you do not know it, or you would have gotten a black, and you do not know it. But this is a bad inference. There is nothing to be known that you don't know. The sense in which you do not know what would have happened is just this: it is not the case that you would have gotten a red and you know it, and it is not the case that you would have gotten a black and you know it. This is the sense in which you do not know the details of a story whose details have never been filled in, the sense in which you do not know the name of Cinderella's birth-mother.

explicit facts of the story. And we can say that it wasn't 100; that too would contradict those facts. But there is presumably some range such that, between that range, we want to say that there is no fact of the matter concerning the number of the soldiers. But now the point. There is no fact of the matter here, and yet... certain answers to this question might still seem better—*more likely*—than others.

Consider. There are entire online fora devoted to *answering* questions like "How many soldiers fought at the battle of Helm's Deep", and so forth—and this is, of course, just the tip of the iceberg. Consider this statement from the summary of the "Battle of the Pelennor Fields" on *Tolkiengateway.net* (a sort of Wikipedia for the Tolkien universe):

> As for Mordor's losses, again, the size of Sauron's great army is not definitely known. The full host was estimated at perhaps 75,000. The Orcs and Trolls of Sauron made up most of the force, though it is known that there were some 18,000 Haradrim. (The Rohirrim, consisting of 6,000 riders, were "thrice outnumbered by the Haradrim alone".) Almost all of the attackers were slain or routed; though not specifically mentioned, all of the War Elephants were likely killed, along with numerous Trolls, Orcs, and Evil Men.

Let us pause to dwell on this author's claim that, "though not specifically mentioned, all of the War Elephants were likely killed" in the battle. Now here we have a puzzle. The author acknowledges that this particular facet of the battle *is not specifically mentioned* by Tolkien. Nevertheless, the author maintains that it is *likely* that all of the Elephants were killed. But if this matter is not specifically mentioned, then it is plausible (given other relevant facts in this context) that there is simply *no fact of the matter* concerning whether all of the Elephants were killed.[11] But if there is no fact of the matter concerning whether all of the Elephants were killed, how could the claim that all of those Elephants were killed be *likely*?

Consider the following. The author of this passage, in all likelihood, would be prepared to grant that there is no fact of the matter about whether, in the fiction, all of the Elephants were killed, or instead whether one escaped in the fray. In other words, if two Tolkien fans (perhaps forgetting the details of the episode in question) got into an (unlikely) debate about whether any of the Elephants escaped—one maintaining that they all perished, the other maintaining that at least one got away—presumably the author of this entry will not maintain that

[11] Note: I am not committing myself to the view that the fiction-determining facts are limited simply to what is "specifically mentioned" in the story; the point here is just that this is a case where (i) the fate of the Elephants is not specifically mentioned, and (ii) there don't seem to be any *other* facts that could be brought to bear here that require the result that they all die or require the result that at least one got away. Thus, this is very plausibly a case where the fiction is "open" or "incomplete". And yet: it does seem likely—in some sense!—that all the Elephants would have died in a battle of this kind, War Elephants being as conspicuous as they are, and the battle being as vicious as it was. But still: strictly speaking, this is something about which there is no fact of the matter.

there are some facts by reference to which one of these parties could be seen to be *wrong*, although we just don't know what they are. Presumably the author of the entry would, at some point, simply invoke the claim that *it wasn't specified* whether any of the Elephants got away; thus, though a case can be made that they all perished (that outcome would seem to be most consonant with the facts of the story), a case could also be made that perhaps one got away, and there is no fact of the matter about whether *as a matter of fact* one got away or instead they all died. And yet: this apparently has not stopped this author from, *in some sense*, siding with the claim that they all died. (In this author's opinion, the case that they all died is the stronger case than the case that maybe one got away.) If I may speak for the author, this author thinks that it is likely that all the Elephants died in the battle—see his or her statement above!—but *not* likely that "In the Battle of Pelennor Fields, all of the War Elephants died" represents a truth (that is, after all, something about which there is no fact of the matter). Regardless of what this author endorses, however, such a position does seem reasonable. The result is that we *seem* to be committed to the truth of the following:

> There is no fact of the matter whether all the War Elephants were killed in the Battle of Pelennor Fields, but all the War Elephants were probably killed in the Battle of Pelennor Fields.

And this can certainly seem puzzling. My purpose here, however, is not fully to resolve what we should say about this puzzling state of affairs; my purpose is to make a comparison with the doctrine of the open future. And the important comparison is the following. When we are thinking that a good case can be made that all of the Elephants perished, we are, in some sense, thinking that their perishing is most consistent with the facts of the story. In other words, *relative to the fiction-determining facts*, the hypothesis that they all died is more likely than the hypothesis that at least one escaped. In some sense, in the most natural way of filling out the story *from the base-level fiction-determining facts*, we reach the result that all the Elephants die. Still, the claim that all the Elephants died can be *made probable* or *likely* in *this* sense, even if there is no fact of the matter about whether they all died—indeed, even if it is *false* that in the fiction, they all died, and *false* that, in the fiction, they all did *not* die. Similarly, *relative to the future-determining facts* (viz., the present and the laws), the hypothesis that there will be rain tomorrow can be more likely than the hypothesis that there will be no rain tomorrow. Still, the claim that there will be rain tomorrow can be *made probable* or *likely* in *this* sense, even if there is no fact of the matter about whether there will be rain tomorrow—indeed, even if it is *false* that there will be rain tomorrow, and *false* that there will be no rain tomorrow.

It should be noted, on this last point, that, as you might expect, I do not think anything in the facts about fictions forces us to deny either bivalence or Excluded Middle. Letting **InF** be an "In the fiction" operator, in the relevant cases (where the fiction is, we might say, "open"), we have ~**InF**p and ~**InF**~p. For instance, we have ~**InF**(The War Elephants all died in the battle), and ~**InF**(The War Elephants did not all die in the battle). And it is worthwhile to observe that, when it comes to *fictions*, the relevant "scopelessness" principle (and the associated principle, say, "Fiction Excluded Middle") would seem plainly to be false. In particular, ~**InF**p certainly does not seem equivalent to **InF**~p, and "**InF**p ∨ **InF**~p" would seem to be a terrible principle—a principle which after all demands that fictions be *completely specific*, which, of course, they aren't.[12] In these respects, the logic of **InF** matches my account of the logic of **F**. However, there are dissimilarities as well. Note that—so long as time is assumed to continue for at least n more units of time—we will always, on my account, have Fn(p ∨ ~p). Now, whereas for *most* fictions we have **InF**(p ∨ ~p), there is no reason why we must; indeed, we can easily imagine *fictions* about what happens when the Law of Non-Contradiction suddenly fails. (Explosion!) Thus, although (p ∨ ~p) certainly may hold *in reality*, it may *not* hold in any given fiction (that is, under the scope of the **InF** operator). To summarize: my suggestion is that the logic of **InF** and the logic of **F** are in these ways similar, *except to the extent that fictions needn't be consistent.* The unfolding present and the laws write the story of the future, but unlike authors of fictions, the present and the laws are constrained to write something *consistent*: they cannot write the story in such a way that we ever get Fn(p ∧ ~p). And though they may currently have left some details of the story unwritten—so that, as of now, we have ~Fnp and ~Fn~p—unlike authors of fictions, the authors of the future must eventually write a completely detailed story. *But the story of the future is not yet written.*

6.13 Unifying the Three Cases

We can now attempt to give a unified account of the problem of probabilities for future contingents, counterfactuals, and fictions. (I am not entirely sure

[12] Well, not so fast. Perhaps unsurprisingly, there are ways of motivating Fiction Excluded Middle; indeed, inspired by Woodward (2012: 785–789), one might motivate it in response to what he calls the *incompleteness problem* for fictionalist strategies in metaphysics. As Woodward brings out, there is an intimate connection between Fiction Excluded Middle and Counterfactual Excluded Middle on at least one reasonable way of making sense of the fictionalist's "According to the fiction..." prefix. Simplifying: if one thinks that, "According to the fiction, p" should be understood as something like, "If the fiction were true, then it would be the case that p", then the adoption of CEM would force the adoption of FEM. Ultimately, one might go supervaluational: where we had thought the fiction was "incomplete", neither **InF**p nor **InF**~p is true, but the disjunction is indeed true. As Woodward notes, this strategy mandates accepting the view that *the fiction* (e.g., "The Lord of the Rings saga") is *referentially indeterminate*: there are various candidate *completions* of the fiction, and it is *indeterminate* to which such completion *the fiction* refers—although it does refer to some such completion. (Cf. once more the models of Chapter 2.) So, sure. In the Lord of the Rings, all the Elephants were killed, or, in the Lord of the Rings, not all the Elephants were killed. It is just indeterminate which—and it is just indeterminate to which completion of the fiction we are referring when we talk about "The Lord of the Rings". What we have here, again, is a deep parallel to debates about future contingents.

about this attempted unification, especially as regards counterfactuals; at any rate, I do think it is worth thinking through whether the following comparisons hold.) Consider fictions first. The problem is that "All of the War Elephants were likely killed" (as asserted in the context mentioned above) can be represented neither as

Pr:**InF**:[All the Elephants are killed]
Nor
InF:**Pr**:[All the Elephants are killed]

The first reading is in conflict with the observation that it is *not the case* that, in the fiction, all the Elephants are killed. Since (*inter alia*) the fiction does not *specify* whether all the Elephants are killed, it is not the case that, in the fiction, all the Elephants are killed. (Again, this is taken to be a matter about which the fiction is "open".) And the second reading is simply the wrong reading. The second reading is a reading which says that, in the fiction of the Lord of the Rings saga, it is probable that all the of Elephants in the Battle of Pelennor Fields were killed. But this is simply false. The Lord of the Rings saga is simply silent about this probability. At any rate, *this* claim must be distinguished at logical form from a *different* sort of claim. Indeed, if we are looking for a claim regarding the Lord of the Rings saga which we might fairly regiment as "**InF**:**Pr**:[*p*]", then we might try, "In the Lord of the Rings saga, it is probable that anyone who takes the Ring of Power is eventually corrupted." In other words, it is not story-internal probable that the Elephants were killed, in anything like the way that it is story-internal probable that anyone who takes the Ring of Power is eventually corrupted. That it is (overwhelmingly) probable that anyone who takes the Ring is corrupted is itself *part of the fiction*. (Admittedly, it may be part of the fiction that this is *guaranteed*, but never mind.) However, *that the Elephants were probably killed* is in no parallel way part of the fiction. But our decision to regiment "All the Elephants were likely killed" as "**InF**:**Pr**:[All the Elephants are killed]" leaves us no way to distinguish *that* claim from claims like that, in the Lord of the Rings, it is probable that anyone who takes the Ring is eventually corrupted.

Now consider the case of counterfactuals. This case is admittedly not as clear as the former, but we *might* contend that "If it had been flipped, the slightly heads-biased coin would probably have landed heads" can be represented neither as

Pr(F > H)
Nor
F > **Pr**(H)

The first reading is in conflict with the observation that it is *not the case* that, if flipped, a slightly heads-biased coin would lands heads. Since *reality doesn't decide* which way such a coin *would* land, it is not the case that such a coin *would* land heads (which is not to say that it *wouldn't*). (This is, at any rate, the Lewisian position I wish to defend.) And the second reading is, perhaps, simply the wrong reading. Briefly: if it had been flipped, heads wouldn't be *probable* (or improbable), but instead *actual* or *not actual.* The relevant probability isn't *counterfactual,* but instead somehow inherent in the *actual* makeup of coin. In other words, there is a difference between saying that, under a certain condition, there *would have been a probability* of a certain event, and saying that, under a certain condition, an event *would probably* have happened. But our decision to regiment the given claim as "F > **Pr**(H)" leaves us no way to distinguish such claims at logical form.

Now consider the case of future contingents. According to the denier of WEM—that is, according at least to me—"There will probably be rain tomorrow" can be represented neither as

Pr:Fn[Rain]
Nor
Fn:**Pr**[Rain]

The first reading is in conflict with the observation that *it is not the case that* there will be rain tomorrow. Since reality is *undecided* regarding what it is going to contain tomorrow, it is not the case that there *will* be rain tomorrow (which is not to say that there *won't* be). And the second reading is simply the wrong reading. We aren't saying that *tomorrow* rain will be probable. Indeed, by tomorrow, rain that day will no longer be "probable" or "improbable"; it will be happening or failing to happen. We are saying that rain is *today* probable *for tomorrow.*

The result is the following. When it comes to "will probably" sentences, my contention is that we cannot semantically "separate out" the *will* and the *probably* in terms of any simple scope distinction: the *probably* cannot be given wide scope over the *will,* and nor can it be given narrow scope with respect to *will*—that is, it cannot be rendered as a *will be probable. Will probably* then, is neither of those two things, but some third thing. In a sense, then, it seems that I am committed to the view that *will probably* is its own semantic unit, as it were. Is this desperately ad hoc? Frankly, I am unsure—but my claim has been that parallel issues arise in the case of fictions and counterfactuals. In the case of fictions: the *probably* (in the relevant cases) can be given neither wide scope over the *in the fiction* operator, and nor can it scope *under* that operator; it is neither of those two things, but some third thing. And so similarly for counterfactuals: in the relevant cases, the *probably* cannot take scope over the whole conditional, and nor can it attach solely to the consequent of such a conditional. It is neither of those two things, but some third thing. What are these alleged third things? I wish I could say. But it is not in my gift to say.[13]

6.14 Conclusion

Well, where are we? I genuinely wish I knew. About all I claim to know is that there is *no easy victory* over the open futurist from observations about betting

[13] With apologies to the reader uninterested in the sorts of horrible complexities that plague the literature on conditionals, which are horrible: my contention here is parallel in important respects to Lewis' treatment of "might counterfactuals", as in "If it were the case the *p*, it might be the case that *q*". (Note that *probably* and *might* certainly seem very similar; *prima facie*, one is just "high chance", and the other "some chance".) Intuitively, it would seem that there are two options for the scope of the *might*: the might scopes over a whole *would* conditional, or the *might* scopes just over the consequent of such a conditional. That is, we might try:

$Might(p > q)$ (Stalnaker 1981: 98–99)
$p > Might\ (q)$ (Bennett 2003: 191)

Lewis contends, however, that neither reading is satisfactory—that is to say, the form of the might counterfactual is neither of those two things, but some third thing (Lewis 1973: 80–81). According to Lewis, this "third thing" is the "not would not" analysis: $\sim(p > \sim q)$ (Lewis 1973). I am not hereby committed to the "not would not" analysis of the might counterfactual—the point is simply that, just as someone attracted to Lewis' approach to the counterfactual will have to give some analysis of the might counterfactual that is neither of the two noted above, plausibly someone attracted to that approach will have to give some analysis of the "would probably" counterfactual that is neither $\mathbf{Pr}(p > q)$ nor $p > \mathbf{Pr}(q)$. And if we can make sense of *that*—and, again, I hope we can—then we will be able to make corresponding sense of "will probably" statements on which they are neither **Pr**:**F**n*p* nor **F**n**Pr**:*p*. And so similarly for the relevant probabilities for fictions. Again: the three cases must be treated in parallel fashion. Notably, Bennett observes that

> Stalnaker observes that Lewis explained 'If... might...' as though it were an idiom, something to be understood as a single linguistic lump, like 'under way': you wouldn't try to explain 'At 5 PM the ship got under way' by explaining 'under' and explaining 'way'. (2003: 189)

And this is akin to the result I am seemingly stuck with about *will probably*. If I had to take a stab at the truth conditions for "It will probably be in n units of time that *p*", however, I would perhaps try something like this: in all of the most likely worlds, in *n* units of time, *p*. That is to say: we must help ourselves to some idea about which worlds of those that are available are most likely; "It will probably be that *p*" is true just in case *p* holds in those worlds. Which worlds are those that are "most likely" or "most probable"? I suggest: those towards the actualization of which the present is currently *tending*. These sorts of analyses, however, have a way of never working. (I am grateful to Brian Rabern for suggesting this comparison with the *might* counterfactual, and for invaluable discussion.)

One final addition. In personal correspondence, Rabern writes:

> Kratzer's (1986) proposal that, in general, if-clauses are restrictors of modals avoids the charge that "if... might" is idiomatic. And the same trick can be applied to "if... would probably". Consider a "would probably" sentence: "If the plate had fell, it would probably have shattered". Following Kratzer, it's natural to suggest that the if-clause restricts the modal 'probably', so that the logical form of "If it had been that *p*, it would probably have been that *q*" is: [Pr: *p*] *q*. (See Schwarz 2018 on how to restrict a probability judgement by a proposition in the subjunctive mood.)
>
> Now—as Wolfgang [Schwarz] suggested—one could extend this idea, so that just as we have a restriction on 'probably' in the cases of counterfactual shifting, we likewise have a restriction on 'probably' in cases of *temporal* shifting. That is, we could treat "probably would, if A" and "probably will, in n units" in an analogous fashion. In this way, "It will probably be in n units of time that *p*" has the form [**Pr**: n](*p*). Much here would need to be fleshed out, but this idea seems worth pursuing.

Indeed, this idea seems worth pursuing; however, I must leave this idea for future work.

behavior, on the one hand, and about credence and probability on the other. On the latter front, I cannot pretend in this chapter to have resolved the questions about how the open futurist should think of the interaction of *will* and *probably*—the issues here are tied, I have argued, to extraordinarily difficult further issues about conditionals, probability, and conditional probability. And the debate about *those* issues certainly remains unresolved, and I can offer little in the way of an attempt to resolve them.[14] All I claim to have done is to have complicated what may seem to be the initial impression that the open futurist faces, from these considerations, some uniquely disastrous result. That conclusion, I hope to have shown, is premature; arguably, anyone not committed to Fiction Excluded Middle will face parallel questions, and the same goes for anyone not committed to Counterfactual Excluded Middle.

[14] Cf. Moss 2013, Hájek 2014, Schulz 2017, Schwarz 2018, and Khoo (forthcoming).

7

Future Contingents and the Logic of Temporal Omniscience

*Patrick Todd and Brian Rabern**

At least since Aristotle's famous 'sea-battle' passages in *On Interpretation* 9, some substantial minority of philosophers has been attracted to what we might call the doctrine of the open future. This doctrine maintains that future contingent statements—roughly, statements saying of causally undetermined events that they will happen—are not true.[1] But, *prima facie*, such views seem inconsistent with the following intuition: if something *has* happened, then (looking backwards) it *was* the case that it *would* happen. How can it be that, looking forwards, it isn't true that there will be a sea-battle, while also being true that, looking backwards, it was the case that there would be a sea-battle? This tension forms, in large part, what might be called *the problem of future contingents*.

Some theorists respond to this tension by insisting that one of the intuitions here must simply be denied. For example, so-called *Peirceans* give up the backward-looking intuition, while so-called *Ockhamists* give up the forward-looking intuition (see Prior 1967: 113–135). But a dominant trend in temporal logic and semantic theorizing about future contingents seeks to validate *both* intuitions. Theorists in this tradition—including some

* [This chapter is a joint paper with Brian Rabern, previously published in *Noûs*. Because it is joint work, I hereby duplicate it verbatim, except for the new appendix, which is authored solely by myself. There is thus, unfortunately, at least some repetition below of some material above. However, I do at a few places below insert new editorial notes, in brackets, like this one. Further, I have not tried to make the notation in this chapter consistent with that of the rest of the book—the differences here are minor, and easily observed.]

For comments and discussion, thanks to Michael Rea, Jonathan Kvanvig, John MacFarlane, Robbie Williams, Stephan Torre, Michael Caie, Wes Holliday, Alex Pruss, Bradley Rettler, Chris Tweedt, Sam Levey, Fabrizio Cariani, Neal Tognazzini, David Plunkett, Jacek Wawer, and anonymous referees. We also acknowledge helpful discussions with our colleagues at the University of Edinburgh, including Wolfgang Schwarz, Bryan Pickel, and Anders Schoubye. Earlier versions of this paper were presented at the second Edinburgh Language Workshop (December 2016), the 91st Joint Session of the Aristotelian Society and the Mind Association (Edinburgh, July 2017), and Dartmouth's Sapientia Lecture Series (August 2017)—we thank these audiences for their questions and objections.

[1] Some such views have it that future contingents are neither true nor false; others maintain that they are instead simply false. For the former sort of view, see, e.g., Thomason (1970) and MacFarlane (2003). For the latter "all false" approach, see Hartshorne (1965), and Prior's "Peircean" semantics in his (1967: 128–135); for a different version of this approach, see Todd (2016a) and Todd (2020), and for criticism, see Schoubye and Rabern (2017) and Wawer (2018).

The Open Future: Why Future Contingents are All False. Patrick Todd, Oxford University Press. © Patrick Todd 2021.
DOI: 10.1093/oso/9780192897916.003.0008

interpretations of Aristotle, but paradigmatically, Thomason (1970), as well as more recent developments in Belnap et al. (2001) and MacFarlane (2003, 2014)—have argued that the apparent tension between the intuitions is in fact merely apparent.[2] In short, such philosophers seek to maintain *both* of the following two theses:

Open-future: Future contingents are not true.

Retro-closure: From the fact that something *is* true, it follows that it *was* the case that it *would* be true.[3]

It is well known that reflection on the problem of future contingents has in many ways been inspired by importantly parallel issues regarding divine foreknowledge and indeterminism. Arthur Prior, whose work figures centrally in these debates, was explicitly motivated by the problems of foreknowledge and human freedom, drew inspiration from ancient and medieval discussions of this problem, and formulated various positions regarding future contingents (e.g., "Ockhamism", after William of Ockham) with an explicit eye towards how they might resolve it.[4] The current paper is, in a sense, a continuation of this Priorean project—one he most rigorously pursues in his 1962 paper, "The Formalities of Omniscience".

The combination of Open-future and Retro-closure, though rigorously investigated in temporal logic, has been underexplored in connection with foreknowledge, omniscience, and related issues. Our contention is this: Once we take up this perspective, and ask what accepting both Open-future and Retro-closure predicts about *omniscience*, we'll see that the view harbors some substantial unnoticed costs. We will argue that a temporal semantics that adopts this conjunction, in fact, rules out the existence of an omniscient being (under certain plausible assumptions)—or, at least, requires that any indeterministic universe lacked an omniscient being at some point in its past. Not only does this prove far too much, we will also argue that the resulting picture, in itself, seems incoherent. Notably, although we will use God as our proxy for certain epistemic ideals, the

[2] For a sample of other authors in this tradition, see Belnap and Green (1994), Brogaard (2008), Markosian (2013), Strobach (2014), and the discussion in Dummett (1973: 391–400). Certain interpretations of Aristotle also fall within this tradition (cf. Thomason 1970: 281). Of course, these authors do not all pursue this reconciliation strategy in precisely the same way.

[3] Note that 'would', as used here, is not indicating a counterfactual situation. This use of 'would' is the simple past tense of 'will'. For example, if on Monday John says "There will be a sea-battle tomorrow", then on Tuesday we may report John's utterance with "John said there would be a sea-battle today". To avoid potentially distracting connections to *counterfactuals* and the "subjunctive mood", however, one could give an alternative gloss on the Retro-closure principle—such as, "From the fact that something *is* true, it follows that it *was* the case that it was *going to be* true", or, more simply, "Anything that happens was going to happen".

[4] Cf. Hasle (2012), Uckelman (2012), and Øhrstrøm (2019).

considerations we adduce here needn't be viewed through the lens of philosophy of religion. When we theorize about an ideal knower, we are theorizing about what an agent ought to believe. Thus, if the conjunction of Open-future and Retro-closure leads to an unacceptable view of ideally rational belief, this casts doubt on that conjunction.[5]

Our aim in what follows is to more fully unpack the problems raised by omniscience for views that maintain *both* Open-future *and* Retro-closure.

7.1 Open-Closurism

We will first briefly explain the theoretical and formal underpinnings of the Open-future and Retro-closure theses, and explain how one might maintain both. We call the resulting view *Open-closurism*.[6]

Open-closurism accepts the doctrine of the open future: that future contingent statements are not true. Underlying the view is a familiar model of the future. Roughly, that model is this: *indeterminism plus no privileged branch.* [Ed.: this corresponds to either model II or III as described in Chapter 2.] In the context of causal indeterminism, we have various "branches" representing causally possible ways things might go from a given moment, consistently with the past and the laws. Importantly, no one branch is "metaphysically privileged" with regard to the others. Future contingents, however, could only be true if one particular branch *was* so privileged. Future contingents are therefore not true.

Yet, Open-closurism also accepts the Retro-closure principle: anything that does in fact happen always would happen. In order to motivate Retro-closure, theorists often point to standard things we say in various conversational contexts. In particular, if someone makes a prediction, and that prediction in fact comes to pass, we may say something like, "You were right!" And this practice seems to

[5] In this way, our project here is deeply similar to Hawthorne's (2005a), "Vagueness and the Mind of God". Hawthorne asks what certain theories about *vagueness* predict about divine omniscience, thereby testing those theories; in our case, we ask what a given theory about future contingency predicts about divine omniscience. More generally, our project overlaps with themes in Williams (2014), who explores, sometimes via consideration of a God-like agent, which theories of rational belief are best paired with certain accounts of indeterminacy.

[6] Open-closurism is reminiscent of certain interpretations of Aristotle's view on future contingents. Of course, the interpretation of Aristotle on future-tensed statements is complex and controversial (see, e.g., Gaskin 1995), so we will not claim that Aristotle was himself an Open-closurist. Thomason, the *locus classicus* of the Open-closurist view, insists that his picture is in line with Aristotelian themes: "It may also be that the theory presented here in fact coincides with the views of previous philosophers on truth and future tenses. Here, Aristotle is the man who comes first to mind; his 'sea-battle' passage is, at first glance anyway, in very good accord with the modelling of the future tense propounded here" (Thomason 1970: 281). Cf. Dummett (1973: 393–394). [Ed.: Note: for a new interpretation of Aristotle's position in *On Interpretation* 9, see Carter (ms)—who argues that, contra so much philosophical tradition, Aristotle was in fact a proto-Peircean, and is best interpreted as maintaining that future contingents are all false!]

presuppose the validity of the Retro-closure principle.[7] For instance, in support of Retro-closure, MacFarlane writes:

> It seems clear that tomorrow we will know more about which of the various possible future contingencies facing us at present were realized. For example, if it is sunny, we'll look back and say, "Yesterday it was the case that Berkeley would be sunny now". (MacFarlane 2014: 212)

In terms of the tense-logical operators, **P** ("one day ago") and **F** ("one day hence"), the Retro-closure principle amounts to the thesis that every instance of the following schema is true: $[\phi \rightarrow \mathbf{PF}\phi]$.[8]

Now, again, some theorists see a tension between Open-future and Retro-closure, and accordingly adopt one in preference to the other. But Open-closurism maintains *both* by putting forward the following picture. Looking forwards, there is no privileged branch. Accordingly, looking forwards, future contingents, such as "There will be a sea-battle tomorrow" and "There will not be a sea-battle tomorrow", which (letting B stand for "there is a sea-battle") might symbolized as $\mathbf{F}B$ and $\mathbf{F}\neg B$ respectively, are not true. However, looking *backwards*, e.g., from the perspective of a current sea-battle, there is, now, a way things *went* to get us to here; accordingly, in a statement such as "It was the case yesterday that there would be a sea-battle today" (symbolized as $\mathbf{PF}B$) when the past tense operator takes us "back" to a point in the "temporal tree" to evaluate the future tensed statement $\mathbf{F}B$, we *do* at that point have, in some sense, a privileged branch of evaluation, viz., the one we took to get us to back to that point. In short, when we have a simple formula $\mathbf{F}\phi$, with ϕ on some but not all branches, then given that there is no privileged branch, the semantic clauses do not deliver a truth. However, when **F** is embedded under **P**, the semantic theory (in some sense) tells you: go back—but then return from whence you came, and check whether ϕ. And thus, the picture validates Retro-closure.

That's, at least, a helpful metaphorical gloss on the view. The way Open-closurism has actually been implemented model-theoretically is by adopting the *supervaluational* method (Thomason 1970).[9] The overall strategy can be divided into two parts. First, the operators **F** and **P** are treated as purely temporal operators—this is in accord with Ockhamism but opposed to Peirceanism,

[7] Thomason insists that the principle is common sense: "arguments such as 'there is space travel; therefore it was the case that space travel would come about' strike us as valid on logical grounds" (Thomason 1970: 268).

[8] If we adopt Prior's metric tense operators (1957: 11–12), $\mathbf{P}_n\phi$ stands for "It was n units of time ago that ϕ", and "$\mathbf{F}_n\phi$" stands for "It will be n units of time hence that ϕ". Note that throughout we will simplify things by using the metric tense operators "one day hence" and "one day ago", and we will abuse notation slightly by using **F** and **P** (instead of $\mathbf{F}_1$ and $\mathbf{P}_1$) for these respectively.

[9] For early developments of supervaluational semantics in application to other cases where truth-value gaps might arise, see Mehlberg (1958: 256–259) and van Fraasen (1966).

where the latter assumes that **F** quantifies over possible worlds in addition to future times. So, the first part of the strategy says that for any world history h and any time t on that history, the satisfaction of **F**ϕ and **P**ϕ by h at t (for any sentence ϕ) are defined as follows:

- **F**ϕ is satisfied by a history h at time t iff ϕ occurs at $t + 1$ on h
- **P**ϕ is satisfied by a history h at time t iff ϕ occurs at t - 1 on h

These clauses specify how the temporal operators "shift" forward and backwards on a given possible history of the world.

Saying this much only specifies when a sentence is *satisfied by a world history at a time*, but it doesn't yet specify when a sentence is *true* at a given moment. Specifying this is the key supervaluational aspect of the Open-closurist approach. Consider all the possible total world histories. Since the view holds that the future is open—indeterminism with no privileged history—a moment might take place on many overlapping world histories, where overlapping histories share a past and laws up to that point, but diverge thereafter. In contrast to the Ockhamist, who insists that a sentence is true just in case it is satisfied by the *privileged* history, the Open-closurist holds that since no history in the overlap is privileged, a sentence is true just in case it is satisfied by *all* the overlapping histories.[10]

Truth: ψ is *true* at a time t iff ψ is satisfied by all histories h that overlap at t, and ψ is *false* at a time t iff ψ is unsatisfied by all histories h that overlap at t, and ψ is *indeterminate* otherwise.

This model supports both Open-future and Retro-closure. Consider the picture in Figure 7.1.

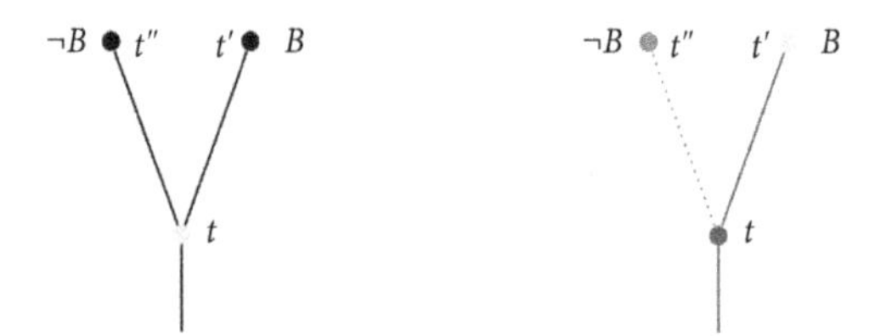

Figure 7.1 Branching

[10] Notice that in this sense the supervaluational method is reminiscent of Tarski's (1935) landmark definition of "truth" in terms of satisfaction by all assignments of values to variables. Tarski restricts the definition of truth to closed formulas, but a nearby definition goes as follows: For any formula ϕ, ϕ is true iff ϕ is satisfied by all sequences, and ϕ is false iff ϕ is unsatisfied by all sequences. On this definition an open formula such as (Gx ∨ ¬Gx) is true, even though neither disjunct is.

Both **F***B* and **F**¬*B* are not true at *t*, since some future histories from that time feature a sea-battle and some don't. (Consult left figure.) But from the perspective of a future time *t'* at which there is a sea-battle, since *B* is true, **PF***B* must also be true: If *B* is true at *t'*, then every history that overlaps at *t'* has a past that has a future that features *B*, so it follows that **PF***B* is also true at *t'*. (Consult the right figure.) In general, ϕ will imply **PF**ϕ, in accordance with the intuitions supporting Retro-closure, and yet we still maintain Open-future. This is the elegant Open-closurist package, which promises a resolution to the Aristotelian puzzles surrounding future contingents.

Such is the formal model of future contingency underlying Open-closurism. To foreshadow what is to come, it is worth observing what sorts of (informal) "dialogues" concerning anticipation and retrospective assessment this model predicts to be perfectly coherent. Suppose Jones believes that there will be a sea-battle tomorrow. Now consider the following dialogue (Dialogue-1):

A: Does Jones correctly believe that there will be a sea-battle tomorrow?
B: It is not true that he does.
A: Does Jones *incorrectly* believe that there will be a sea-battle tomorrow?
B: It is not true that he does.
A: So, the future is open?
B: Precisely. It is indeterminate whether Jones' belief is correct.
[*...a day passes, and a sea-battle rages*]
A: Did Jones correctly believe yesterday that there would be a sea-battle today?
B: Yes, of course he did. He believed that there would be a sea-battle today—and there is a sea-battle today.

The position of the Open-closurist is that B's pattern of response is perfectly coherent, and furthermore, could be perfectly accurate. And now note what seems to be the consequence of the accuracy of B's position: the past would seem to have undergone a sort of *change*. Crucially, however, it has undergone merely what we might call an *extrinsic* change—or a so-called "Cambridge change". More particularly, in the dialogue, we have "moved" (over time) from the untruth of "Jones' belief is correct" to the later truth of "Jones' belief was correct". Thus: at a certain point in time, it is not true that Jones' belief has a certain property (the property of being correct). Later, however, Jones' belief did have that property at that time.

At this stage, however, it is important to note that the proponent of Open-closure will insist that this sort of "change in the past" is not the sort of *radical* "change in the past" which clearly seems impossible. For instance, suppose that, on a given day, "Jones is in Los Angeles" is untrue, but then, on the next day, "Jones was in Los Angeles yesterday" is true—or, in another preview of what's to come, consider the move from the initial untruth today of "Jones believes that ϕ" to the later truth of "Jones believed that ϕ yesterday". Intuitively, *these* sorts of "changes" would require *intrinsic* changes in the past—and these sorts of changes, the Open-closurist can insist, are the ones that are impossible. (More about these issues shortly.)

However, the change at issue in the dialogue above is *not* a change of this kind. For consider: whether a given belief counts as being *correct* or *incorrect* would plainly seem to be a *relational* property of that belief; whether a belief is correct or incorrect is constituted, roughly, by how that belief is *related* to the world. Thus, in the dialogue above, when a sea-battle comes to pass, this brings it about that Jones' prior belief was correct (when he held it). However, had a sea-battle *failed* to come to pass (which was objectively possible), this would have brought it about that Jones' prior belief was *in*correct (when he held it). However, it is crucial to observe that in *both* scenarios, "the past"—in the ordinary sense of "the past"—is exactly the same: the difference is solely that, in one scenario, a past belief comes to have had a certain relational property, and in the other scenario, that belief comes to have had a *different* (incompatible) relational property. The past, however, remains *intrinsically* just the same in both scenarios.

As we will see, these differences—between *intrinsic* and *extrinsic* changes in the past—play a crucial role in our arguments to come.[11]

7.2 The Logic of Temporal Omniscience

Our contention is that Open-closurism predicts certain problematic consequences regarding the logic of divine omniscience. The important connections between the logics of tense and divine omniscience are often noted in the literature on future contingents. For example, the following passage from Peter Øhrstrøm and Per Hasle provides a nice point of departure:

[11] As we note below (in fn. 18), the changes required by the Open-closurist are changes in what have been called the *soft* facts about the past. (For an introduction to this distinction, see the essays in Fischer 1989, and Todd and Fischer's more recent survey in their 2015.) But isn't it widely accepted that changes in the soft facts about the past are perfectly admissible? No—or, better, that depends. What has been widely accepted is that we can act in ways that would *require* such changes. But there is an enormous (and crucial) difference between the following two theses: (1) we can, but never do, act in ways that would require Cambridge-changes in the past, and (2) we can, and often do, Cambridge-change the past. (Cf. Todd and Fischer 2015: 13.) And whereas the truth of (1) is widely accepted (in the literature on fatalism and free will), it is the truth of (2) that is at issue for the Open-closurist. Compare: (3) we can, but never do, act in ways that would require that the facts about the future would be different, and (4) we can, and often do, change the facts about the future. Whereas (3) is widely accepted, the only theorist ever to accept (4) was Peter Geach—a more or less unknown position he developed in his (1977: Ch. 3). (For more recent developments of this "mutable futurist" approach, see Todd 2011, 2016b.) In short, we do not mean to precipitously concede to the Open-closurist that the requisite Cambridge-changes in the past are perfectly acceptable; we mean only to concede, for the moment, that they are *less* unacceptable than the parallel *intrinsic* changes.

> The medieval discussion regarding the logic of divine foreknowledge is, from a formal point of view, very close to the classical discussion concerning future contingency. If we add the assumption that necessarily, something is true if and only if it is known to God, then it is easy to see how the discussion regarding the logic of divine foreknowledge is, from a formal point of view, essentially the same discussion as the classical discussion concerning future contingency. This was clearly realised by the medieval logicians. (Øhrstrøm and Hasle 2011)

The formal equivalence suggested by Øhrstrøm and Hasle could be developed in different ways, but in what follows, we develop it primarily in terms of constraints on the *beliefs* of an omniscient being: *God believes all and only what is the case.* This slogan, however, could be cashed out in at least two competing ways. [Ed.: Here we reintroduce the two pictures of "omniscience" discussed in the previous chapter.] The first way to capture the slogan is in terms of an intuitive principle we will call *Omni-accuracy.*

Omni-accuracy: ϕ if and only if God believes ϕ

We will argue that this principle combined with Open-closurism quickly leads to some undesirable results.

While some Open-closurists may happily accept Omni-accuracy and insist that the consequences we draw out are not so undesirable, some will presumably insist on an alternative rendering of the intuitive slogan. In the context of supervaluationism, the Open-closurist will want some means of distinguishing "It is *true* that ϕ" from "ϕ". That is, letting T be an object language operator expressing "truth", the Open-closurist rejects the following equivalence: ϕ iff $T\phi$.[12] This opens up space for a second, and non-equivalent, principle connecting God's beliefs to what is the case, namely *Omni-correctness.*

Omni-correctness: $T\phi$ if and only if God believes ϕ

These, then, are the two options characterizing (a necessary condition on) divine omniscience that we will explore in connection with Open-closurism.[13]

[12] There are choices here as to how to define the truth predicate, and we are not insisting that this is the only kind of truth predicate available to the supervaluationist. We are only assuming that the supervaluationist will want to make the relevant distinction somehow, and we are providing them with T as the way to make that distinction: $T\phi$ is satisfied by a history h at time t iff ϕ is satisfied by every history h' overlapping at t. Note that Thomason (1970: 278) instead introduces a "transparent" truth predicate: $T\phi$ is satisfied by a history h at time t iff ϕ is satisfied by h at t. See also MacFarlane (2014: 93).

[13] Note that the conception of God we are working with in this paper is one in which God exists *in time*, not "outside of time" (cf. Prior 1962: 116). Within the philosophy of religion, there are two conceptions of "divine eternity": one on which God is *sempiternal* (exists at all times) and one on which God is atemporally eternal (exists outside of time). Here we assume sempiternalism; God's omniscience

For ease of exposition, we will often talk in terms of *God's anticipations* and *God's recollections*.[14] We assume that for God to believe that something will happen tomorrow just is for God to anticipate it. And for God to believe that something happened yesterday just is for God to remember it. So, letting '**Bel**', '**Ant**', and '**Rem**' be divine belief, anticipation, and remembrance operators, respectively, we will often employ the locutions on the right-hand-side of the following equivalencies:

- $\mathbf{Bel}\ F\phi \leftrightarrow \mathbf{Ant}\ \phi$
- $\mathbf{Bel}\ P\phi \leftrightarrow \mathbf{Rem}\ \phi$

With these abbreviations we can also contrast Omni-accuracy and Omni-correctness as follows.[15]

Option 1. Omni-accuracy:	*Option 2.* Omni-correctness:
$\phi \leftrightarrow \mathbf{Bel}\ \phi$	$T\phi \leftrightarrow \mathbf{Bel}\ \phi$
$F\phi \leftrightarrow \mathbf{Ant}\ \phi$	$TF\phi \leftrightarrow \mathbf{Ant}\ \phi$
$P\phi \leftrightarrow \mathbf{Rem}\ \phi$	$TP\phi \leftrightarrow \mathbf{Rem}\ \phi$

Given the validity of one set of these principles one can substitute equivalents and preserve truth. For example, according to option 1, it follows that:

$$FP\phi \leftrightarrow \mathbf{Ant}(\mathbf{Rem}\ \phi)$$

And thus combined with a principle of tense-logic such as $[\phi \rightarrow FP\phi]$, we have:

$$\phi \rightarrow \mathbf{Ant}(\mathbf{Rem}\ \phi)$$

More naturally: if ϕ, then God anticipates remembering that ϕ. For example: if a sea-battle is ongoing, then God anticipates remembering the sea-battle. The principle captures a natural thought: anything that happens will always be remembered by God.

Now, we could, of course, detain ourselves for some time developing the parallels between various principles in tense-logic with their "theological"

is *temporal* omniscience. For a classic discussion of these issues, see Stump and Kretzmann (1981); see further Pike (1970) and Leftow (1991).

[14] Strictly speaking, we are talking about what God *seems* to remember—or God's *apparent* memories. "Remembering that..." is arguably factive, so one can't remember an event that didn't take place. But for God any *apparent* (or "quasi") memory is also accurate.

[15] To be clear, just as with the Retro-closure principle, the claim here is not merely that these biconditionals are true; it is that the schemata are valid in the sense that they hold for any sentence ϕ and for all worlds and all times. This strong equivalence vindicates the substitution.

counterparts; we believe that these parallels deserve a more thorough treatment than that which we propose to give them in this paper. (On this approach, we transform the logic of the tenses into the logic of divine anticipations and remembrances.) But we now have enough on the table to assess the two options, given the assumptions of both Open-future *and* Retro-closure.

7.3 The Costs of Omni-Accuracy

To cut to the chase, consider what is, according to option 1, the theological counterpart of the Retro-closure principle [$\phi \rightarrow \mathbf{PF}\phi$], viz.:

$\phi \rightarrow \mathbf{Rem}(\mathbf{Ant}\ \phi)$

More naturally: if ϕ, then God remembers anticipating that ϕ. For example: if there is a sea-battle (ongoing), then God remembers anticipating that sea-battle yesterday. More simply: if there is a sea-battle today, then yesterday God anticipated a sea-battle today. Now, here we have a principle with direct and obvious implications for the traditional picture of divine foreknowledge—and a principle whose implications have been debated for millennia. From the fact that something has happened, does it follow that God has always anticipated it? This is, of course, the traditional, orthodox position on divine foreknowledge, and this implication would certainly be accepted by contemporary proponents of such orthodoxy (e.g., Plantinga 1986)—and it certainly would have been accepted by Ockham. Indeed, the principle arguably encapsulates precisely the spirit of Ockham—and other defenders of the traditional picture of divine foreknowledge. When Augustine complains (in *On Free Choice of the Will*) that it would be absurd to deny that God has foreknowledge, precisely his complaint is that it would be absurd to maintain that there are things that happen which God hasn't always known (viz., anticipated) would happen.

Such a principle, of course, has its defenders, and its attractions (both theological and otherwise). But such a principle seems plainly to be in tension with the doctrine of the open future. The tension might be brought out my means of the following dialogue (Dialogue-2):

US: God, do you anticipate a sea-battle tomorrow?
GOD: It is not true that I do.
US: Do you anticipate peace tomorrow?
GOD: It is not true that I do.
US: So, the future is open?
GOD: Precisely.
[...*a day passes, and a sea-battle rages*]

US: God, did you anticipate this sea-battle?
GOD: Yes, of course I did.

But surely this is unacceptable. How can this make sense, unless God has fundamentally changed the past? According to the Omni-accuracy principle, the open future licenses God's initial claim that it is not true that he has the anticipation. When God faces the open future—and sees that things could go so that B or so that $\neg B$—it is not true that **F**B, so it is not true that God anticipates that B. But likewise, Retro-closure licenses God's maintaining that he had the anticipation all along: Retro-closure plus Omni-accuracy yields that *everything has been anticipated by God*. There is thus a challenge for the Open-closurist who accepts Omni-accuracy: they must explain how it is that God could have the set of seemingly impossible attitudes exemplified in Dialogue-2.

But let's slow down. Recall the issues at the end of section 7.1: Open-closurism requires the coherence of *extrinsic* or "mere Cambridge" changes in the past. As we saw, it requires a "move" (over time) from the untruth of "Jones' belief is correct" to the later truth of "Jones' belief was correct." But we distinguished *that* sort of "change in the past" with a different sort of change in the past: an *intrinsic* change in the past—the sort of change in the past that more clearly seems objectionable. And now the problem: the sort of change in the past involved in Dialogue-2 would seem to imply an *intrinsic* change in the past; we have moved from the initial untruth of "God believes that ϕ" to the later truth of "God believed that ϕ". We do not profess to know the operations of the divine mind. But we do claim that if this move represents those operations, those operations imply an intrinsic change in the past.

We can thus represent our argument against the conjunction of Open-closurism and Omni-accuracy slightly more carefully as follows:

1. If Open-closurism and Omni-accuracy, then God's combination of attitudes in Dialogue-2 are possible.
2. The attitudes exemplified in Dialogue-2 necessarily imply an intrinsic change in the past.
3. Intrinsic changes in the past are impossible. So,
4. It is not the case that: Open-closurism and Omni-accuracy.

We have brought out how Open-closurism together with Omni-accuracy predicts the pattern of response in Dialogue-2, and we have thereby defended (1). In this paper, we simply assume the truth of (3).[16] That leaves (2). Might the Open-closurist insist that, on closer inspection, the "move" at issue in Dialogue-2

[16] We think this assumption is dialectically reasonable, and we certainly are not aware of any place our interlocutors in this essay (e.g., Thomason or MacFarlane) have denied it.

implies no more of an intrinsic change in the past than the "move" at issue in Dialogue-1?

Recall: the Open-closurist under consideration accepts the view that it is not true that there will be a sea-battle tomorrow, and it is not true that there will be peace tomorrow—but they *also* accept that it is not false that there will be a sea-battle tomorrow, and not false that there will be peace tomorrow. Whether there will be a battle tomorrow is strictly *indeterminate:* $\neg \mathrm{TF}B$ and $\neg \mathrm{T}\neg \mathrm{F}B$. Thus, crucially, given Omni-accuracy, they must also accept:

Unsettled Mind: For some ϕ, $\neg \mathrm{T}(\mathbf{Ant}\ \phi)$, and $\neg \mathrm{T}(\neg \mathbf{Ant}\ \phi)$.

That is, given that it is indeterminate whether there will be a sea-battle, it is also indeterminate whether God anticipates a sea-battle.[17] But if it is indeterminate whether God anticipates a sea-battle, then perhaps we can say the following: God's mind is either in a state of sea-battle-anticipation or it's in a state of non-anticipation, but it is metaphysically indeterminate which. And if we can say *that*, then perhaps we can also say that the coming to pass of a sea-battle retroactively constitutes the (prior) state of God's mind as having been the anticipation of a sea-battle. Prior to the sea-battle, no one (not even God!) can tell determinately whether the relevant mental state is the anticipation of a sea-battle (because it is *not* determinately such an anticipation). But once the sea-battle transpires, God's mental state *had been* (all along) the anticipation of a sea-battle. Thus, in an important sense, what we do *now* partially constitutes which mental state God had been in—the belief-state that we would battle, or instead the belief-state that we would not battle. Thus, the changes at issue concerning God's mental state would be mere *extrinsic* changes on analogy with the sorts of changes already acknowledged to be required for the Open-closurist's treatment of future contingents. And if this is so, premise (2) is false.

This, then, is the picture that the proponent of Open-closurism and Omni-accuracy must defend. Such a picture is, of course, mysterious—but we think it's even worse than that. Consider the nature of the "indeterminacy" of God's belief-states that this approach must posit. God's beliefs concerning future contingents are indeterminate in the sense that *what belief-state* God is in constitutively depends on what eventuates in the future—that is, constitutively depends on whether or not a sea-battle eventuates. However, very plausibly, whether God currently counts as believing that there will be a sea-battle tomorrow doesn't await the unfolding of time. Nevertheless, this is what the view under consideration

[17] One might find independent support for this stance on God's mind in Caie (2012). Caie argues that if ϕ is indeterminate, then a rational agent ought to be such that it is indeterminate whether he or she believes that ϕ. Thus, it would follow that when God, a perfectly rational agent, faces the open future, it is indeterminate what beliefs God has about the future.

must be insisting: *Whether God counts as having a certain present anticipation constitutively depends on what the future has not yet settled.*[18] This sort of indeterminacy, which we will call "future history indeterminacy", can be defined as follows:

Definition. ϕ is *future history indeterminate* at t iff there are some possible histories overlapping at t according to which ϕ and some possible histories overlapping at t according to which $\neg\phi$. (And ϕ is *future history determinate* otherwise.)[19]

Intuitively, however, whether someone counts as believing that an event will happen is not indeterminate, in this sense. That is, *belief* and *anticipation* would seem to be future-history determinate affairs: whether a person has or lacks a given

[18] Readers familiar with the literature on divine foreknowledge (especially in the wake of Pike 1965) and the associated "hard/soft fact" distinction may recognize this position; essentially this position has been defended by Zemach and Widerker (1988):

> For all we know, the fact that p may be an environmental necessary condition for the internal state of God, m, to count as the belief that p. It may be that m is God's belief that p only if p is the case, and thus he who is able to bring it about that not-p is able to bring it about that m is not a belief that p. (Zemach and Widerker 1988, in Fischer 1989: 118)

They elaborate:

> The fact that p does not *cause* God's mental state m to mean 'p'; rather, it is *in virtue of* its being the case that p, that God's mental state m means 'p'. Thus, the property *is a belief that* p is a relational property m has in virtue of its relation to the fact that p. (Ibid.: 119)

In consequence,

> It is not that through our action we can bring about the non-occurrence of an event in the past. Rather, through our action we can deprive a past event from having a certain relational property, a property which accrues to it by virtue of the occurrence of a certain future event over which we have control. Since, as argued above, God's belief that Q is a relation obtaining between a certain mental state of God m and the fact that Q, we can, by exercising our control over the latter, bring it about that the mental state would, or would not, count as a belief that Q. (Ibid.: 121)

Thus:

> It is indeed sometimes within our power to determine what God believes. We do not thereby cause any changes in God, nor limit His omniscience, for it is neither change nor limitation in God that some of His states count as beliefs of what we do in virtue of our doing those very things. (Ibid.: 122)

And here we have a position that maintains precisely what we have just wished to deny. But our complaint against such a picture is the same as John Martin Fischer's (1994: 120–125). According to Fischer, it is, *inter alia*, extremely difficult to see how any such picture can plausibly maintain that God indeed has *beliefs*. A full discussion of this position must lay outside the scope of the present paper. Briefly, however, our main contention is that, on this view, God does not genuinely have *beliefs* in the first place; God may have *geliefs* (where whether one *gelieves* that something will happen is partly a matter of whether it will happen), but not *beliefs*. Thus, the adoption of this radical position on God's mind is *not* a way of vindicating the pattern of response in Dialogue-2 (wherein God has genuine *beliefs* about the future)—and thus the adoption of this position is no response to our argument in this paper. Note: because (necessarily) someone believes p iff p, it does not follow, by itself, that whether that person *believes* p constitutively depends, in the noted sense, on whether p.

[19] This notion of a future-history determinate statement is essentially the notion of a "moment-determinate" affair as defined in Belnap and Green (1994: 374) and MacFarlane (2014: 214). The intuitive idea is that a moment-determinate affair doesn't constitutively depend on the unsettled future.

belief at t does not depend, in this sense, on what happens in the future relative to t.[20] Notice that, in this respect, *belief* differs importantly from *correct belief*. As we brought out in Dialogue-1, whether one counts as *correctly believing* that an event will take place is, at least in part, a matter of (is constitutively dependent on) whether in fact it will take place. Contrary to the current suggestion, however, whether one counts as *believing* that an event will take place is *not* constitutively dependent on whether it will take place. And so this way of denying premise (2) seems untenable.

An Open-closurist may be tempted at this point to just dig in, and accept the radical idea that God's anticipations are constitutively dependent on the future, in the same way that *correct belief* is constitutively dependent on the future. But it is not enough just to accept the consequence that some *anticipations work in mysterious ways*. The indeterminacy in God's mind will tend to bleed out. God's beliefs may co-vary with other affairs that one would be hard pressed to accept as future-history indeterminate affairs. Consider, for example, God's *actions*. Current actions or utterances would seem to be good examples of future-history determinate affairs, and God's actions are linked to his beliefs.[21] Assuming that God can *act* on the basis of his beliefs about the future, the tension that arises can be brought out in the following (Dialogue-3):[22]

US: Do you anticipate a sea-battle next year?
GOD: It is not true that I do.
US: What would be rational for you to do, if you did anticipate a sea-battle next year?
GOD: I would employ 1000 workers from Tyre to take those stones in the quarry to construct a wall around the city.
US: And peace?
GOD: I would employ 1000 workers from Sidon to take those same stones and instead construct a temple in the center of the city.
US: Are you currently doing either of those things?

[20] Here we are in agreement with Belnap and Green, who insist that "whether a person asserts (wonders, hopes, bets) [and, we might add, believes] that A does not depend upon what history has not yet settled" (Belnap and Green 1994: 382). Note: here we are plainly discussing the central themes at issue in the so-called "hard"/"soft" fact distinction in debates about foreknowledge and free will. For a defense of this characterization of God's beliefs (as temporally future-non-relational, "hard" facts at times), see Todd (2013a, 2013b). For more on these issues, see the essays in Fischer (1989), and Todd and Fischer (2015).

[21] The argument here doesn't rely on God's actions *per se*. This is just an illustration. We just require some future-history determinate witness for the following: *If God anticipates a sea-battle, then some future-history determinate fact obtains that would not obtain if God did not anticipate a sea-battle.*

[22] Note: there are well-known difficulties associated with the idea of God acting on the basis of such beliefs. But these difficulties arise only on *Ockhamist* assumptions about those beliefs (viz., that they are both infallible and *comprehensive*). (See, for instance, Hasker 1989: 53–64, Hunt 1993, and Robinson 2004.) The openness of the future, however, removes these difficulties, since these beliefs will not be comprehensive—and so it would seem ad hoc to deny that God could act on the basis of his beliefs about the future.

One possibility at this stage is for God simply to say *no*: the indeterminacy of his anticipations does not extend to the indeterminacy of his *plans*. Such a position, combined with Retro-closure, encounters a severe version of the difficulty to be noted shortly. So instead suppose God says:

> GOD: It is not true that I am, nor true that I am not.

Such a posture is, of course, difficult to comprehend. God maintains that it is neither true nor false that he is employing 1000 workers from Tyre to build a wall using some given stones, and also neither true nor false that he is employing 1000 different workers to use those same stones instead to build a temple. He is doing one or the other, but it is metaphysically indeterminate which. Needless to say, this is puzzling. (For instance: what does the city look like right now?) But this is not all. For on either such approach, we get a problem like the following:

> [*...a year passes, and a sea-battle rages*]
> US: Did you anticipate a sea-battle a year ago?
> GOD: Yes, I did.
> US: Then why didn't you employ those 1000 workers from Tyre to construct a wall around the city? The rampaging army will be here soon!

Needless to say, such a question seems reasonable. How does God respond according to this characterization? Does God say:

> GOD: What? Behold: there indeed have been workers from Tyre building such a wall with those stones over the past year; haven't you noticed the influx of Tyronians? Fear not: the wall is in good stead (and this is why there is no temple in the center of the city). After all: I anticipated this sea-battle.
> US [DUMBSTRUCK]: Oh my God, look at the wall!

But surely this is unacceptable. For suppose that, instead of the commander declaring war, that commander had instead commanded peace. Then God would have instead had to say:

> GOD: What? Behold: there indeed have been 1000 Sidonians in the city using those stones to build a temple in the center of the city (that is, after all, why there is no wall around the city). Worry not: the temple is in good stead. After all: I anticipated precisely this peace.

And it is fundamentally unclear how one and the same set of circumstances could resolve itself into the correctness of *both* of these speeches: if we get war, then God will be able to make the first speech, and if we get peace, God will be able to make the second. This seems unacceptable—if not simply impossible. The reason these situations strike us as impossible is that affairs such as an agent's current actions or utterances or the current physical locations of stones are future-history determinate affairs. But if such affairs are linked to God's indeterminate anticipations, then they would also have to be indeterminate—but they aren't. Needless to say, these dialogues raise a great many questions, not all of which we address. We simply note the following: it is unclear how they could have adequate answers.

7.4 The Costs of Omni-Correctness

Open-closurism combined with Omni-accuracy has led to some undesirable results. But as we mentioned at the outset, this is not the only way that one might try to cash out the slogan, *God believes all and only what is the case.* Some Open-closurists will no doubt insist on an alternative rendering of the slogan. *Truth*, they will say, is satisfaction by every overlapping history, and thus the truth predicate should be defined as follows:

- Tϕ is satisfied by a history h at time t iff ϕ is satisfied by every history h' overlapping at t.

Given this understanding the following equivalence must be rejected: ϕ iff Tϕ. And those who reject this equivalence will naturally insist on a principle connecting God's beliefs to what is *true* (cf. Dummett 1973: 398):

Omni-correctness: Tϕ if and only if God believes ϕ

If this is the constraint God is working under, then we must imagine God responding very differently in the dialogue (Dialogue-4):

US: Do you anticipate a sea-battle tomorrow?
GOD: No.
US: Why not?
GOD: Well, it isn't *true* that there will be a sea-battle tomorrow. The future is open.
US: So there are no *truths* that escape your gaze?
GOD: Correct.
US: And in that sense you are omniscient?
GOD: Correct.
[... *a day passes, and a sea-battle rages*]
US: God, did you anticipate the sea-battle?

Now, what Open-closurism plus Omni-correctness predicts is *not* that God will maintain that he had anticipated the sea-battle. This approach instead predicts the following:

GOD: Well... no. I didn't anticipate the sea-battle.

And here God is certainly being consistent. But now we continue as follows:

US: But a sea-battle was going to occur! [**PF***B*]
GOD: Granted.
US: So something was going to happen that you didn't anticipate would happen. [$\mathbf{P}(\mathbf{F}B \wedge \neg\mathbf{Ant}\ B)$]
GOD: Granted.
US: But isn't that just to say that you weren't omniscient after all?
GOD: ...

Now, as a first approximation, the problem is that we seem to have shown that God was not omniscient. After all, God seems to be admitting former ignorance. If there are events that were indeed always going to happen that God didn't anticipate would happen, then in what sense was God omniscient? Given the principle of Omni-correctness and the Open-closurist model, the following statement is true at the sea-battle: $\mathbf{P}(\mathbf{F}B \wedge \neg\mathbf{Ant}\ B)$. Thus, some instances of the schema $\mathbf{P}(\phi \wedge \neg\mathbf{Bel}\ \phi)$ are true. Normally, one would take a true instance of that schema to be a statement to the effect that *God was ignorant*: Something was the case that God didn't believe was the case!

Now in response to this complaint, one might maintain that God is not and was not genuinely "ignorant", since one is ignorant only if there is a *truth* about which that one is ignorant. However, according to the view under discussion, there was no *truth* of which God was ignorant. At the time of the sea-battle, there was always going to be a sea-battle, but it wasn't always *true* that there would be a sea-battle. That is, since *B* is true, then **PF***B* is also true, but **PTF***B* isn't. So while it is right that the sea-battle was going to happen and God didn't anticipate it—$\mathbf{P}(\mathbf{F}B \wedge \neg\mathbf{Ant}\ B)$—there is nevertheless no *truth* that escaped his gaze, since it wasn't *true* that there would be a sea-battle—$\neg\mathbf{PTF}B$.[23]

[23] Notice that the view also predicts that the following disjunction is true (determinately, super-, true): $(\mathbf{F}B \wedge \neg\mathbf{Ant}\ B) \vee (\mathbf{F}\neg B \wedge \neg\mathbf{Ant}\ \neg B)$. That is, either there will be a sea-battle tomorrow and God doesn't anticipate the sea-battle *or* there will be peace tomorrow and God doesn't anticipate peace. Thus, it would seem, something is the case that God doesn't believe is the case. Again, normally, one would take that to be a statement to the effect that *God is ignorant*. But then why call such a being "omniscient"? The response again is this: God is not currently ignorant, since there is no *truth* about what is going to happen that God fails to anticipate, since both $\neg\mathbf{TF}B$ and $\neg\mathbf{TF}\neg B$. Hawthorne (2005a) has suggested that supervaluationism applied to *vagueness* has an analogous result for an omniscient being: Either (Frank is bald and God doesn't know it) or (Frank is not bald and God doesn't know that). And he insists that the supervaluationist can perhaps learn to live with

We think that one can accept this view only at the expense of giving up on the fundamental intuitions that motivate Retro-closure in the first place. Very plausibly, if one is moved by the backward-looking intuition that, given that a sea-battle has occurred, it was always going to occur, it seems that one should likewise be moved by the intuition that given that a sea-battle has occurred, it was always *true*—which is not to say determined!—that it was going to occur. However, by treating *truth* as, in effect, synonymous with *determined*, the view under consideration makes it impossible to express the intuition that, though it was true that the sea-battle would occur, it wasn't *determined* that it would occur. This is, however, an intuition we should be able to express—and this is precisely the intuition that motivates Retro-closure.

Notice that MacFarlane, the archetypical Open-closurist, agrees with this latter intuition, and it is, in fact, what motivates him to sophisticate the supervaluationist picture by adding on a kind of truth *relativism*. Here is a characteristic passage:

> According to supervaluationism, then, my utterance was not true. By [the definition of **T** above], the sentence I uttered was neither true nor false at the context in which I uttered it. But surely that is the wrong verdict. I said that it would be sunny today, and look—it is sunny! How could it be, then, that what I said was not true? To see how strange the supervaluationist's verdict is, suppose that the Director of the Bureau of Quantum Weather Prediction now offers me an irrefutable proof that, at the time of my utterance yesterday, it was still an open possibility that it would not be sunny today. Would such a proof compel me to withdraw my assertion? Hardly. If I had asserted that it was settled that it would be sunny today, I would have to stand corrected. But I did not assert that. I just said that it would be sunny—and it is. My prediction was true, as we can demonstrate simply by looking outside. (MacFarlane 2014: 89–90)

Of course, in this passage, MacFarlane is not suggesting that we give up Open-future. He thinks we need to vindicate both the claim that future contingents are neither true nor false, and the retrospective assessments that some future

this result given that they already tolerate the following: Either (Frank is bald and it is not true that Frank is bald) or (Frank is not bald and it is not true that Frank is not bald). But the case of future contingents adds an important complicating factor, which makes this line of thought less appealing. The indeterminacy involved with the future involves a dynamic aspect that has no analogue with respect to vagueness—in the vagueness case, there is no "waiting around" to see how the indeterminacy gets resolved (so that we can *then* say that it was the former: Frank was, indeed, bald, but God didn't know that). That is, there is no principle that is analogous to the Retro-closure principle. And so whereas we may be able to accept "Either (Frank is bald and God doesn't know that) or (Frank is not bald and God doesn't know that)", it is substantially more difficult to accept the (backwards-looking) discharged disjunct, $\mathbf{P}(\phi \wedge \neg \mathbf{Bel}\ \phi)$.

contingents were true. MacFarlane presents the tension as the following puzzle: *present claims concerning the future can be shown to be untrue by a proof of present unsettledness, but past claims concerning the present cannot be shown to have been untrue by a proof of past unsettledness* (MacFarlane 2014: 90). What the puzzle motivates is a conception of truth that validates both of the following principles (using a generic truth predicate T^*):

Retro-closure: For all ϕ, $\phi \rightarrow \mathbf{PT^*F}\phi$

Open-future: For some ϕ, $(\neg T^*F\phi \wedge \neg T^*F\neg\phi)$

But validating the latter would seem to invalidate the former—the forward-looking intuition seems to require a robust notion of truth which quantifies over histories, whereas the backward-looking intuition seems to require a more-or-less *transparent* notion of truth.[24] MacFarlane insists that we should "split the difference" by introducing a definition of truth with "double time references"—the time of utterance and the time of assessment (MacFarlane 2003: 331; cf. Dummett 1973: 394–395).[25] Various technicalities can be employed at this point to vindicate both principles. But this is not our primary concern. Our point, instead, is this: insofar as the Open-closurist view has a notion of truth that vindicates the (updated) Retro-closure principle, they will have to accept the conclusion that *God was genuinely ignorant*. Something was *true* (in the relevant sense) that God didn't believe: $\mathbf{P}(T^*\mathbf{F}B \wedge \neg\mathbf{Ant}\ B)$. This is a conclusion MacFarlane must simply accept (on the assumption that MacFarlane does not wish to accept the first option, Omni-accuracy). In other words: MacFarlane is right about the supervaluationist. But we are right about MacFarlane. On his picture, God was ignorant. The question now becomes: is this result defensible? More particularly, is it (1) defensible that a theory of temporal semantics alone could rule out the former existence of an omniscient being in an indeterministic universe? And (2) is it *plausible* that, given the open future, we can nevertheless fairly charge God with having been ignorant—as Open-closurism suggests? It is these questions we take up in the remainder of the paper.

[24] Roughly, a notion of truth is "transparent" just in case it predicts no difference in "ϕ" and "It is true that ϕ."

[25] MacFarlane tends to only talk about relativistic truth in the metalanguage, where he says, e.g., "F*B*" is not true at *t* assessed from *t* and "F¬*B*" is not true at *t* assessed from *t*, but "F*B*" was true at *t* as assessed from the sea-battle at *t'* (2014: 226). Although MacFarlane employs this talk of relativistic truth in the metalanguage, he doesn't actually introduce an operator T^* which corresponds to the metalanguage. In fact, the only truth predicate MacFarlane introduces in the object language is what he calls *monadic truth* (2014: 93–94). The monadic truth predicate "True" is *transparent* in the sense that the following equivalence holds: True ϕ iff ϕ. But then, in this sense, it is false that future contingents are not True. We will set monadic truth aside. All that really matters here is that MacFarlane somehow wants to vindicate that backward-looking claims that it was true that a sea-battle would occur.

7.5 Ruling Out Omniscience?

Given that Open-closurism has the implications for omniscience we have outlined above, it seems that one could argue from Open-closurism and indeterminism to a substantial metaphysical conclusion:

1. Open-closurism is the correct semantic theory of temporal language.
2. The universe is indeterministic.
3. If Open-closurism is correct and the universe is indeterministic, then at some past time the universe lacked an omniscient being.
4. Therefore, at some past time the universe lacked an omniscient being.

Now, as a first approximation, the problem here is that this seems to prove too much. Needless to say, we are not insisting that since there indeed has always been an omniscient being in our indeterministic universe, and since the Open-closurist must deny that this is so, Open-closurism is false. Of course, our results do point to the following: theists—that is, those who *do* believe that there exists and has always existed an omniscient being—plausibly should not be Open-closurists. And that is certainly an interesting, important result in itself. The point we wish to make is instead the following. Just as it is not for the semanticist to say whether the future is causally open, it is likewise not for the semanticist to say whether the universe contains or ever did contain an omniscient being. Here we are arguably following the advice of MacFarlane himself:

> A proper account of the semantics of future contingents can vindicate ordinary thought and talk about the future in a way that is compatible with branching. [...] we assume neither that physical law is deterministic nor that it is not. That is a question for physics. Semantics, conceived as a theory of linguistic meaning, should not presuppose any particular answer to this question. The project is not to give a semantics for future-directed talk that assumes indeterminism, but rather to give one that does not assume determinism.
>
> (MacFarlane 2014: 202–204)

Nor, we think, should a semantics for future-directed talk make presuppositions about the existence or non-existence of an omniscient being. This is a question for the metaphysician, or perhaps the philosopher of religion, or perhaps even the person in the pew—but at any rate it is not a question for the semanticist *qua* semanticist. In general, one could argue that a semantic theory—a theory concerned with the logic and compositional structure of the language—ought not settle certain substantive non-semantic questions. Although we find it very attractive, we can't hope to offer a defense of this general semantic neutrality principle here.[26] But the appeal to neutrality we are making is much narrower in scope: A correct semantic theory for temporal language must be compatible with the existence of an omniscient agent in a (deterministic or indeterministic) universe.

It is worth observing that the main alternative views concerning the semantics for future contingents don't fail to be neutral in this way. Clearly, the Peircean can maintain the claim that, yesterday, there existed an omniscient God; the Peircean, in virtue of denying Retro-closure, will simply contend that, though yesterday God did not anticipate today's sea-battle, this doesn't show that yesterday God was ignorant—for, according to the Peircean, yesterday it wasn't *true* that there would be a sea-battle today. Similarly, the Ockhamist can plainly maintain that there exists and did exist an omniscient being (witness, for instance, Ockham). At any rate, if there is no Ockhamist *God* (no being that knows or did know the Ockhamist facts, as it were), this is certainly not the fault of the Ockhamist *semantics*. But the Open-closurist semanticist—in virtue of being such a semanticist—*cannot* maintain the claim (in the relevant context) that, yesterday, there existed an omniscient being. In this, the Open-closurists stand alone—and problematically so.

To flesh out this complaint, it is useful to compare the Open-closurist view with a view that might initially be seen as a partner in crime—that is, with a nearby view that *also* denies that there was an omniscient being, but does so on roughly *metaphysical* rather than semantic grounds. In particular, consider the picture endorsed by certain so-called "open theists" such as Swinburne, Hasker, and van Inwagen.[27] Like Open-closurists, such theists accept the thesis that past indeterminism implies that God was ignorant.[28] According to this version of open theism, that is, it was true that certain events were going to happen which God had not anticipated would happen. However, the central argument these

[26] The appeal to semantic neutrality is not novel with us. Above we quoted MacFarlane (2014) in connection to the neutrality of temporal semantics on determinism/indeterminism, but others have made similar appeals in other domains. For example, Yalcin (2010) is concerned with the semantics of the language of probability, and maintains neutrality on the metaphysical issues concerning "interpretations of probability". He says, "We will consider natural language as we find it, without making assumptions about the nature of the domain(s) being described in advance.... As we will see, one can make considerable progress limning the logic and compositional structure of probability operators in abstraction from substantive metaphysical assumptions" (917). Likewise, Cariani (2014) defends the thesis that a semantic theory for normative language should be neutral between a range of normative and evaluative theories, and uses this to argue against certain theories of deontic modals that are not neutral in this regard. However, for complications concerning the neutrality constraint, see Cariani and Santorio (2018: 144).

[27] Swinburne (2016: 175–199), Hasker (1989), van Inwagen (2008).

[28] However, they do not accept this result under this description. Instead, they seek to argue that, since the given truths were impossible to know, God can still be called "omniscient", despite not knowing them. These philosophers argue that, just as omnipotence requires only an ability to do what is logically possible to do, omniscience only requires knowledge of what is logically possible to know. We disagree: if there are truths that a being doesn't know, that being is not properly called omniscient, *even if* those truths are impossible to know (Kvanvig 1986: 14–25). The better option for these philosophers is simply to deny that God is omniscient, but to maintain that God is as perfect knower as there could be anyhow. However, we set this complicating factor aside.

philosophers make at this stage is that it was *impossible*, even for a perfect knower, to anticipate these events, even though it was true that they were going to happen. *Prima facie*, Open-closurists might make exactly the same appeal: it was true that the events were going to happen—but anticipating them was impossible, even for a perfect knower.

The crucial difference between the given version of open theism and the view of the Open-closurist, however, concerns the proffered *grounds* of this impossibility. For the Open-closurist, it is semantic—whereas for the open theist, it is metaphysical. More particularly, on this open theist view, we have the following (Dialogue-5):

US: Do you anticipate a sea-battle tomorrow?
GOD: No.
US: Why not?
GOD: Well, for all I know, it is *true* that there will be a sea-battle tomorrow—it is just that, supposing that this is true, it is a truth I am not in position to know.
US: So there are truths that escape your gaze?
GOD: Correct.
US: And in that sense you are ignorant?
GOD: Correct.
[...*time passes, and a sea-battle rages*]
US: God, did you anticipate this sea-battle yesterday?
GOD: As I said yesterday, I didn't believe that there would be a sea-battle today—although now we can see that it was *true* that there would be a sea-battle today.
US: Well, what's your excuse? I thought you were meant to be omniscient.
GOD: So some people say; but I am not, and was not. Let me explain. Yesterday it was *true* that there would be a sea-battle today—but this wasn't *determined*. Accordingly, there is nothing that I could have "looked at" yesterday to verify that there would be a sea-battle today. I could have known that there would be a sea-battle today only if I had some mystical insight into the contingent truths about the future—but (contrary to the well-meaning suggestions of my friend Plantinga[29]) no one has or could have any such mystical insight.
US: So your excuse for not believing that there would be a sea-battle today was solely the excuse of *non-determination*, and not the excuse of non-truth.
GOD: Correct.

This view, then, simply *denies* Open-future (some future contingents are just true), but *accepts* that God doesn't anticipate the truths about the contingent

[29] Plantinga (1993).

future. Now, such a position may or may not be adequate, and its costs have been well-documented already.[30] The important point, for our purposes, is the grounds this view offers for the non-existence of an omniscient being. And the point is that those grounds are metaphysical, not semantic. Semantically, such open theists are *Ockhamists* (some future contingents are simply true), and there is no motivation from the Ockhamist semantics, per se, toward the rejection of an omniscient being. For example, on this view, although it is causally possible that there will be sea-battle tomorrow and causally possible that there won't be, according to the Ockhamist there in fact (e.g.) will be a sea-battle tomorrow. And if God is omniscient, then God anticipates tomorrow's (non-determined) sea-battle. That's coherent. But these open theists reject omniscience because they insist that no one—not even God—could have access to the facts about the contingent future. So, what is telling these philosophers that there was no omniscient being is (very broadly speaking) their metaphysics of mind. ("No one has or could have any such mystical insight.") It is not their semantic theory by itself.

7.6 Revoking Omniscience

Our first complaint against Open-closurism (combined with Omni-correctness) is that such a theory, in itself, predicts whether and when an indeterministic universe contained an omniscient being. But even if we take on board these strong commitments, the resulting model of the ideal knower has the following implausible feature: The title of "knower of all the truths" is retrospectively revoked at each passing moment. We now turn to this second complaint.

Recall the position of the open theists discussed above. According to this view, if we ask the ideal knower—God—during the sea-battle whether he had been ignorant of the sea-battle, he will of course admit that he was. But on this view God simply *starts* by admitting that he is ignorant, and so it is hardly a mystery that retrospectively God should likewise admit that he *had been* ignorant.

The model provided by Open-closurism is importantly different on this front. God needn't admit current ignorance. Indeed, God should *deny* current ignorance, precisely in virtue of maintaining that the future is open. Looking forward into the future, that is, God has the excuse of non-truth: it isn't true that there will be a sea-battle tomorrow, and *that* is why he doesn't believe there will be a sea-

[30] On this approach, we have what has been called an "Ockhamist" tense-logic (for a defense of which see Rosenkranz 2012), but we do not employ it for purposes that would have pleased Ockham. Instead, though there is a "thin red line" marking a privileged branch, its location is inaccessible even to God. For a critical discussion of this version of open theism, see Todd (2014). We set aside the seemingly remarkable opposite view—attributed to Peter Auriol (ca. 1280–1322)—that though the future is open, in the sense that there are no truths concerning the contingent future, God nevertheless has anticipations concerning the contingent future (Schabel 2000, Knuuttila 2014).

battle tomorrow. Retrospectively, however, God does *not* have the excuse of non-truth. That is, in virtue of granting Retro-closure, when God looks back on the previous day, he is forced to admit that he *had been* ignorant. Given that God is fully aware of this impending revocation of his good epistemic status, it seems that he would be trapped in a perplexing cycle of self-doubt. To draw this out, consider this variant on Dialogue-4:

US: Do you anticipate a sea-battle tomorrow? Or do you anticipate peace tomorrow?

GOD: Neither. It isn't true that there will be a sea-battle tomorrow, but nor is it true that there will be peace. The future is open.

US: But there are no *truths* that escape your gaze? And in that sense, you are omniscient?

GOD: Correct. I am omniscient.

US: Yet, either there will be a sea-battle tomorrow or there will be peace tomorrow. Right?

GOD: Right.

US: So tomorrow you will either be saying "It was true that a sea-battle would occur" or you will be saying "It was true that no sea-battle would occur."

GOD: That's right.

US: So, you will be making one of those statements while recognizing that you anticipated neither a battle nor peace.

GOD: Right.

US: So tomorrow you will either be admitting "It was true that a sea-battle would occur but I didn't anticipate that" or admitting "It was true that no sea-battle would occur but I didn't anticipate that".

GOD: Yes, since no truth escapes my gaze, that is what I foresee: I'll be saying "Yesterday some truth escaped my gaze".

US: So, there is no truth that escapes your gaze, but tomorrow you will admit that some truth *did* in fact escape your gaze.

GOD: Yes.

US: So why insist that you are omniscient if your future self will insist that you weren't?

GOD: [voice inaudible]

God seems to be flouting a sort of reflection principle: you shouldn't believe something if you think your future-self will disagree. The Open-closurist model predicts that at a given time the ideal knower is omniscient (and the ideal knower believes this), while the ideal knower nevertheless foresees that his omniscience will be revoked. This is a mysterious feature of the model. Plausibly, however, if God counts as being omniscient at a time, then it would seem to be a once-and-for-all assessment that God counts as having that feature at that time. But on the model under consideration, whether God counts as being *omniscient at a given time* depends on the temporal perspective. Relative to *today* God is omniscient today, but relative to *tomorrow* God is not omniscient today.

One might be tempted to insist that this is just the mystery of relativism at work. But this reply is inadequate. The sort of relativism at issue is explicitly motivated by (and *only* by) our (alleged) intuitive verdicts about what it is correct to say concerning future contingency and retrospective assessment. And what we have brought out is that relativism (and Open-closurism more generally) does *not* accord with our intuitive assessment of what is correct to say in these domains. We do not find it correct to suppose that, though omniscient today, it could nevertheless be that tomorrow the ideally rational agent will be saying, "Yesterday I had not been omniscient." More to the point, whereas the Open-closurist is right that we *do* find Open-future intuitive, and we *do* find Retro-closure intuitive, we have brought out that we do *not* find the consequences of the *conjunction* of these claims intuitive.

The intuitive incompatibility of Open-future and Retro-closure can be summed up as follows: If there is a robust intuition that if the future is open, then God can—contrary to what God grants in Dialogue-4—deny past ignorance, then there is a robust intuition that if Open-future is true, then Retro-closure is not. Here, then, we must at last bring out the plausibility of God's simply *denying* Retro-closure, precisely on grounds of the open future. If we *begin* once more by granting that God is omniscient, despite not believing that there will be a sea-battle (and not believing that there will not be a sea-battle), our contention is that God's response intuitively should be different:

US: But it was true that a sea-battle was going to occur! And so something was true that you didn't believe! And so: you were ignorant.
GOD: Well, wait. Recall: previously you had *granted* to me that I wasn't ignorant. These were the words out of your mouth: "You are omniscient." Weren't they?
US: Yes.
GOD: But now you're trying to tell me that I *was* ignorant?
US: Yes.

And this seems odd. At this point, it seems that God should maintain the following:

> God: Well, I deny the charge. Just because a sea-battle *did* occur, this doesn't imply that it was going to occur–and so even though a sea-battle occurred, and I didn't believe that a sea-battle would occur, it doesn't follow that I was not omniscient.

And what we have here is God simply denying Retro-closure. Now, our point is not that the denial of Retro-closure *in itself* is plausible, or unproblematic. Our contention instead is that, *in the context of this dialogue*, God has a point. In the context of an admission that the future is open, God should maintain the following: just because the sea-battle occurred, this doesn't imply that it was true that it would occur. And so what we have, in effect, is a way of motivating the following thought: if you grant Open-future, you should deny Retro-closure. Otherwise, God would lack the point he evidently does seem to have.

7.7 Conclusion

The problem of future contingents has traditionally been connected to parallel issues regarding divine foreknowledge, and we have taken up this perspective in order to spell out what a temporal semantics that accepts both Open-future *and* Retro-closure predicts about omniscience. We've argued that the resulting Open-closurist model has substantial unnoticed costs. The Open-closurist cannot maintain the classical view that God is Omni-accurate without accepting that God's anticipations are implausibly constitutively dependent on the future. But the more promising position for Open-closurist, which abandons Omni-accuracy in favor of Omni-correctness, implausibly predicts, by itself, that there was no formerly omniscient being in an indeterministic universe, and encounters the startling result that God had been ignorant—despite the openness of the future! In light of these results, perhaps Open Futurists should resume the recently much neglected project, not of explaining how they might *save* Retro-closure, but how they might credibly deny it.

Appendix: Denying Retro-Closure

My co-author and I ended the above article with a suggestion: Open Futurists should explain how we might credibly *deny* Retro-closure. That is precisely the suggestion that I wish now to take up.

Let me begin with the following distinction. On the one hand, there is an *argument* we have already considered for the validity of Retro-closure—an argument that its validity is required to vindicate our practices of *betting*. Recall Prior's concern, addressed in Chapter 6, that unless Retro-closure is granted, if someone bets on Phar Lap to win a race, and then Phar Lap does win, we might still refuse to grant the payout. I have responded at length to this concern in Chapter 6; Retro-closure, I contend, is in no way required to vindicate our ordinary practices of betting.[31] But besides this sort of argument

[31] It is of course worth noting that, despite Prior's articulation of the relevant worry about betting, Prior's deeper conviction was that Retro-closure should be denied:

> One of the big differences between the past and the future is that once something has become past, it is, as it were, out of our reach—once a thing has happened, nothing we can

for Retro-closure, we do sometimes encounter in the literature expressions of the sentiment that Retro-closure is *directly intuitively obvious.* Consider, for instance, Thomason's contention, noted above, that

> arguments such as 'there is space travel; therefore it was the case that space travel would come about' strike us as valid on logical grounds. (Thomason 1970: 268)

But this won't do. I do not have a theory of "validity on logical grounds," but I do contend that a principle that is "valid on logical grounds" should give us valid inferences at the first moment of time; but suppose—more than a bit fancifully—that there is space travel at the first moment of time. Then the inference that it *was* the case at some *earlier* time that space travel would come about is certainly a bad inference. (More plausible examples could be supplied.) At any rate, consider the modified claim that

> arguments such as 'there is space travel; and there was a long span of time before space travel; therefore anyone during that span of time who didn't anticipate that space travel would come about was not omniscient' strike us as valid on logical grounds.

But do they? I am prepared to admit that this inference will strike many as "valid" on *theoretical* grounds. At the same time, is it completely obvious that, given that space travel came about, anyone who didn't anticipate its coming about (long in advance) was not omniscient? Well, this is obvious only if it is obvious that, long ago, it was *true* that space travel would come about. But why suppose that, long ago, there was any such truth to be known?

Notably, those in the literature who have written in support of Retro-closure typically do so by appeal, not to the principle directly, but to our assessments of imagined *assertions* or *predictions.* And here my approach to Retro-closure builds on ideas articulated in the 1950s by Richard Taylor. Taylor explains the problem as follows:

> Suppose someone, "A," indulged in prophecy, asserting, "Henry will sneeze tomorrow," and another person, "B," following Aristotle's principles, replied, "No, he might, or he might not; it cannot yet be true either that he will or that he will not, this being in the realm of contingencies." Tomorrow comes, and Henry sneezes. A, it would seem, can now say, "I said he would sneeze, and he did, so what I said was true, while you, in denying that what I said was true, are now shown to have been wrong." This comment by A seems reasonable, for it certainly seems that yesterday A had something that B did not have—namely, a true opinion. Of course, B did not say Henry would not sneeze, but still, his opinion was not as good as A's—for A's opinion, we now discover, was true, while B's was just noncommittal.

Such is the problem for the open futurist. How does Taylor respond?

> The most this argument can be claimed to prove is that either A's prophecy was true or that it <u>became</u> true, just as it became fulfilled, through the lapse of time and the reduction to zero of alternative possibilities. There is nothing in it to show that it was

> do can make it not to have happened. But the future is to some extent, even though it is only to a very small extent, something we can make for ourselves. [...] In my own logic with tenses I would express it this way: we can lay it down as a law that whatever now is the case will always have been the case; but we can't interchange past and future here and lay it down that whatever now is the case has always been going to be the case—I don't think that's a logical law at all; for if something is the work of a free agent, then it wasn't going to be the case until that agent decided that it was. ("Some Free Thinking about Time," printed in Copeland 1996: 47–48)

> antecedently true, any more than that it was antecedently fulfilled. Or, to put it otherwise, all the argument shows is the trivial fact that when "tomorrow" had ceased to be tomorrow and had become today, it contained just those events which then happened; it does not show that, on the day before, it was going to contain those rather than alternative ones. No advantage, in the way of true opinion, can be claimed by A as having obtained when he first made his prediction, for all he can claim is that it was fulfilled—which suffices for any wagers that were made. The apparent advantage of his opinion over B's is only an ex post facto sort of one—much like the advantage one might have who, by taking one path rather than another, stumbles upon a fortune. B, on the other hand, has had from the beginning a real advantage, for he claimed the future to be ambiguous and unsettled—as in fact it then was. His opinion, unlike A's, did not have to wait to become true but was true from the start. It only became an inadequate opinion, but not disconfirmed, when A's prediction came true, that is, when the event in question ceased to be a future contingency and to admit of any possibility of being otherwise. [Underlined emphasis added] (1957: 27–28)

I am not entirely sure what Taylor is saying in this passage—but, by and large, I agree with it. However, two brief notes. Taylor seems, in the final two lines, to run together A's *opinion* coming true and A's *prediction* coming true. It is, however, the latter notion only that I wish to investigate and defend. Second, I am inclined to disagree with Taylor that the fact that A's prediction *came true* shows that B's position is or was in any sense "inadequate"; this is a point I thus propose to ignore.

But now the important point. Note Taylor's contention that A's prediction *came true*. It is worth remarking that the language of a "prediction coming true" is utterly ubiquitous and familiar. (Notably, however, this language is almost entirely ignored in debates about the Open Future and Retro-closure.) The crucial question is thus the following: is the Open Futurist who denies Retro-closure entitled to say that the relevant predictions *came true?* More generally, our questions here are twofold: (a) what is it for a prediction to "come true", and (b) can we appropriately say that a prediction "came true" without commitment to Retro-closure?

Prima facie, it is difficult to see why we should not be able to answer (b) in the affirmative. After all, regarding (a), in what does A's prediction "coming true" consist? My answer is the following: A predicted that the event would happen, and it did in fact happen. Accordingly, A's prediction *came true*. More particularly, suppose someone predicts that it will be the case that *p*. Another way to express this claim is that this person predicts that it will become the case that *p*. Accordingly, if someone predicted that it would be the case that *p*, she predicted that it would (later) become the case that *p*. Now, suppose it *is* (that is, does become) the case that *p*. It follows that what she predicted would become true (viz., *p*) did become true. Her prediction thus came true, precisely in the sense that what she predicted would become true (viz., *p*) did become true. No doubt we could pause to develop a considerably more sophisticated formal framework to capture and generalize this basic idea. But we can content ourselves with the following: if anything even approaching this account of "coming true" is on track, then the Open Futurist who denies Retro-closure is perfectly entitled to the claim that the relevant predictions *came true.*[32]

[32] A notable exception to my claim that our language of "coming true" is routinely ignored in these contexts is perhaps Rhoda 2010 (who is replying to a trenchant articulation of the "credence problem" in Pruss 2010). Rhoda does appeal to the idea of "becoming true"—but, for Rhoda, this amounts to future contingent propositions (FCPs) *themselves* becoming true: "It becomes clear that what we have good reason for believing is not that some FCPs *are* true, but rather that some FCPs have a good chance of *becoming* true" (2010: 197). But we must distinguish between the implausible contention that (the

But we can say slightly more. Plausibly, what it is for a prediction to *come true* is for it to be "vindicated" in the sense articulated elsewhere by Green and Perloff and Belnap.[33] The basic idea here is simple. If you make a prediction that it will be that *p*, then any *p*-future is a future in which your prediction is vindicated, and any ~*p*-future is a future in which it is impugned. However, it can nevertheless be *open* right now whether such a prediction is *going to be* vindicated or instead impugned. Note: here we must be careful not to conflate the vindication of *predictions* with the vindication of *predictors*. Your *prediction* may be "vindicated", but *you* may not be vindicated for *making it*; if you make an utterly rash and groundless prediction that nevertheless comes to pass, you can, and do, remain a perfect fool for making it, even if your prediction is vindicated in the sense here identified.

With these notions in mind, we can give the following characterization of predictions "coming true" and being "vindicated" in terms of the framework defended in this book:

> *Classical Open Futurist Coming True*
>
> There are several total ways things may go from here; some have a sea-battle tomorrow, and some do not. In ways things go from here in which there is a sea-battle tomorrow, the prediction that there will be a sea-battle tomorrow *comes true* tomorrow. In that sense, it is *vindicated* tomorrow. In ways things could go from here in which there is no sea-battle tomorrow, the prediction that there will be a sea-battle tomorrow does not come true tomorrow. In that sense, it is *impugned* tomorrow. But there is, now, no fact of the matter concerning whether this prediction will be vindicated tomorrow or will be impugned tomorrow. Consequently, concerning the prediction that there will be a sea-battle tomorrow, it is false that that prediction will come true tomorrow, and false that that prediction will *not* come true tomorrow. Nevertheless, as concerns the prediction that there will be a sea-battle tomorrow, it will be tomorrow that that prediction comes true or does not come true.

What we have here is a simple application of the theory developed in Chapters 2–5. If you make a prediction regarding a future contingent event, then it is not the case that it is going to be vindicated, and not the case that it is going to be impugned—but it is going to be vindicated or impugned. (Once more, we can have ~Fn*p*, ~Fn~*p*, but Fn(*p* ∨ ~*p*).) Now we *wait*. Say it was vindicated. Does it follow that it was *going to be* vindicated? No. In other language: Say that it *came true*. Does it follow that it was *going to* come true? No. And yet: it did.

We are now ready to apply these points to a range of arguments for Retro-closure one can find in the literature. For instance, Alfred Freddoso writes as follows:

> Suppose I predict that on the next toss the coin in your hand will come up heads. And suppose for the sake of argument that the coin's coming up one way or the other is wholly indeterminate, so that prior to the toss the world is not tending, even non-deterministically, either toward the coin's coming up heads or towards its coming up tails .... Suppose, finally, that when you toss the coin, it in fact comes up heads. In that

future contingent itself) "There will be a sea-battle tomorrow" came true today (what would *that* mean? [although I can think of some things it could mean]) and the plausible contention that (the embedded content of the future contingent) "There is a sea-battle" came true today. At any rate, if someone yesterday predicted today's sea-battle, the gloss I would give to "your prediction came true!" is not anything like, "'There will be a sea-battle tomorrow' came true!", but instead "What you said would become true became true!" Cf. also Hasker (1985: 127), who remarks in passing, "We sometimes say of a prediction that it has 'come true', which is not quite the same as saying that it was true all along."

[33] Green (2014: 155), Perloff and Belnap (2012). Note: Perloff and Belnap nevertheless attempt to *retain* Retro-closure—and in this respect our approaches are importantly different.

> case it is perfectly reasonable for me to claim that my prediction was true, that is, that I spoke the truth in asserting beforehand the proposition *The coin will come up heads.* So it is reasonable for me to maintain that this proposition was true before you tossed the coin. (1988: 71)

By way of response: I certainly do not wish to *deny* that it is reasonable to maintain that the proposition *The coin will come up heads* was true prior to the coin toss. (Ockhamism is a reasonable doctrine.) What I instead seek to deny is that it is *unreasonable* to maintain that this proposition *was not* true prior to the coin toss—and it is clearly this stronger thesis that is required in order to construct an argument against the open futurist. More particularly, what is certainly reasonable is that Freddoso's prediction *came true*. That is, it would be unreasonable to deny that Freddoso's prediction was *borne out* or *vindicated*. Further, it would certainly be unreasonable to refuse to grant Freddoso any relevant pay-outs, should any wagers have been placed on the result of the toss. However, as we saw above, open futurism does not make it reasonable to engage in any such refusal, and the open futurist can (and clearly should) grant that Freddoso's prediction *came true*. And this, it seems, is all that reasonability demands in this case.[34]

Consider further this passage from Greg Restall:

> I will give just one argument to the effect that this principle is valid [in Restall's notation: "*p* entails [-][+]*p*"]. Suppose at point *c*, as I stand looking at my tie collection, my son Zachary and my spouse Christine are there, and Zachary says 'Dad will wear a green tie' and Christine says, 'Greg will wear a brown tie'. Then, retrospectively, from the point of view of *g* [in which Restall wears a green tie], what Zachary said at *c* was correct, and what Christine says was incorrect. (2011: 234)

A few moments later, Restall adds:

> If *g* is the case, then any prior prediction—to the effect that it will be the case that *g*—pays off. (2011: 234)

This addition, however, seems to me to effectively *cancel* the argument just given for the claim that *p* implies **PF***p*. As Restall seems to indicate, the intuition here is that any prior prediction of this outcome *pays off*. As we saw in Chapter 6, however, the open futurist can happily grant that any such prediction appropriately "pays off" in this way. Of course, in this case there is an important complication; according to Restall, we do not have a *bet* (a traditional *wager*) "paying off", but instead a prediction itself. But in what sense might a mere prediction "pay off"? (Of course, one may get *credit* in certain social contexts for making a prediction that comes to pass, but it is unclear if this is what Restall has in mind.) In short, if the intuition is that Zachary's prediction was correct (when he made it), then this is an intuition I deny (assuming Restall's choice was indeterministic); if the intuition is that Zachary's prediction came true, this is an intuition we can accommodate; and, finally, if the intuition is that had Zachary *bet* that Restall would wear a green tie, Zachary's bet would pay off, then this too is an intuition we can accommodate.[35]

Now consider this passage from Moruzzi and Wright:

[34] Cf. Green (2014: 156): "The intuitive datum that theorizing in this area must respect is that many predictions eventually are either borne out or not. This, however, is a datum that the Open Future view can accommodate." For a more recent (forthright) denial of Retro-closure, see Green (2020).

[35] Both Freddoso and Restall present their arguments within the context of debates about divine foreknowledge and future contingents. It is fair to say that Freddoso's argument has been enormously influential in philosophy of religion. See further, e.g., Fischer (1989: 27), Flint (1998: 130), and (more recently) Mares and Perszyk (2011: 106–107), all of whom discuss it favorably.

> Of what in fact proves to have transpired it will truly be affirmable that it *was going to* transpire, so that it is true to say that one who predicted the actual course of events spoke truly at the time of her prediction, even though this cannot truly be said before what she predicted "comes true", as we are wont to say.
>
> It is a nice question what drives the Determinacy claim. Certainly it is entrenched in our linguistic habits. I make a prediction. Things turn out as I predicted. And I say, "See, I *was* right." (2009: 317)

It is Moruzzi and Wright who have here italicized the "was": "See, I *was* right." Perhaps Moruzzi and Wright here mean only to draw attention to the past tense construction, which would seem to indicate that the relevant rightness is in the past. It is worth pointing out, however, that in fact we do *not* emphasize the "was" in this way in ordinary speech. You predict that Phar Lap will win. Phar Lap wins. Do you say, "See, I *was* right"—emphasizing the *was*? No—or anyway not unless you know that I am a recalcitrant open futurist and you are trying to make some sort of philosophical point. And if by emphasizing the *was* in this way you *were* trying to make this sort of point—that it follows from Phar Lap's win that he would win—then I would be inclined to deny that it follows from Phar Lap's victory that you "were right" in this sense. (Nevertheless, I would certainly be happy to pay up, if we had *bet* on Phar Lap's victory; you had claimed the Phar Lap-winning futures, and that suffices for that.)

At this stage, we must pause to note a certain sort of ambiguity in saying that someone "was right". Note that, if I make a given prediction, and then things turn out as I predicted, saying, "See, I was right" seems most natural in cases in which I take myself to have had strong warrant for making the prediction that I made. (If someone rashly predicts that the ball will land in slot 17, and then, miraculously, it does, and the person says, in a cock-sure manner, "See! I was right!", we might be inclined to reply that you *got it* right, but that *you* weren't right.) In *this* sense of "was right", however, someone saying, "See, I was right" is not implying the truth of Retro-closure; they are instead claiming that they had *warrant* for their prediction, and implying that the fact that their prediction came true is good evidence that it was likely to come true, precisely as they predicted. And it seems to me that the open futurist can *grant* that, in many such cases, the relevant parties *did* have warrant for their predictions. Clearly, the open futurist can grant that many people have been warranted in *projecting* certain events or *rationally expecting* those events. When an expert makes an entirely warranted projection of a major hurricane next weekend, and then, when that hurricane comes to pass, says, "See, I was right", no open futurist needs to think that he speaks falsely, *even if* it is maintained that "There will be a hurricane next weekend" was not true at the time of his projection. For in saying that he was right, he *may* simply be saying that he was right *to make that projection*—and he was, given the probabilities that obtained at the relevant time, and would have been right to do so *even if* the improbable happened and no hurricane came to pass. Further, in practice, what we call *predictions* often seem to be somehow *weaker* than outright assertions, and are, again, best described as something like *projections*: if someone says, "Looking at the numbers, I'm projecting a major victory for Team A", and then team A wins on precisely the grounds that that person made her projection, then *even the open futurist* should be happy to say (or otherwise concede that) "You were right!" For this may mean nothing more than *you were right to make that projection*.[36]

[36] For more on the norms and nature of *prediction* and its relation to *assertion*, see Benton (2012), Benton and Turri (2014), and Cariani (2020).

But back to the central points at hand. In speaking of the "Determinacy claim," Moruzzi and Wright are following MacFarlane's influential construal of the problem of future contingents as a clash between what he calls the "indeterminacy intuition" and the "determinacy intuition". The indeterminacy intuition is the intuition that future contingents are not true in advance. The determinacy intuition is the intuition that, once the relevant events come to pass, it seems right to say that the given future contingents were true. As we have seen above, according to MacFarlane, an adequate approach to future contingents must respect both intuitions—but respecting both intuitions requires the given sort of relativism about truth. At this stage, however, I wish to call into question whether the "determinacy intuition" needs to be accommodated in the first place. MacFarlane writes that the following

> ... reasoning seems unimpeachable:
> Jake asserted yesterday that there would be a sea battle today
> There is a sea battle today
> So Jake's assertion was true.
> When we take this retrospective view, we are driven to assign a determinate truth-value to Jake's utterance: this is the determinacy intuition. (2003: 325)

Certainly I grant that the Retro-closure principle *makes* such reasoning "unimpeachable"—but what I deny is that the evident unimpeachability of this reasoning decisively supports Retro-closure. What is in fact unimpeachable is not the reasoning displayed above, but instead:

> Jake asserted yesterday that there would be a sea-battle today
> There is a sea-battle today
> So Jake's assertion came true.

—where Jake's assertion "coming true" simply consists in him having asserted that there would be a sea-battle today, and there being a sea-battle today. As I see it, we are liable to confuse the unimpeachability of *this* reasoning with the (theoretically loaded) reasoning articulated by MacFarlane above.

In fairness to MacFarlane, I have defended the claim that we can say that a given *prediction* "came true" without commitment to Retro-closure—but MacFarlane makes his point in terms of *assertion*: "So Jake's *assertion* was true." However, although the language of a *prediction* "coming true" is much more familiar than that of an *assertion* "coming true", a simple Google search reveals that the latter locution is similarly widespread. Herewith several real-world examples culled from the web; more could be provided:

> In the end, President Kennedy's assertion came true, though he didn't live to witness it.
>
> After the Staples volleyball team fell to Darien in the FCIAC quarterfinals last Tuesday, coach Jon Shepro predicted his team would see the Blue Wave one more time in the state tournament. A week later Shepro's assertion came true as the No. 17 Wreckers swept No. 16 Shelton 3-0.
>
> Last night, that assertion came true. Britain is in military lock step with America, as the Bush administration goes to war.
>
> Friends and colleagues of 51-year old Local Government Officer John Stones were in shock today after his much-repeated assertion came true.

But if it is felicitous to maintain that an assertion *came true*—and it is—then it is the reasoning *I* have displayed that is indeed unimpeachable, and it is MacFarlane's that is by comparison questionable. After all, the claim that Stones' assertion *was true* is not obviously equivalent to (or otherwise entailed by) the claim that it *came true*.

Finally, Berit Brogaard writes:

> Suppose, for instance, that you and I are present at *t*2 in the midst of a sea battle. You might say
>
> 2. The sentence 'There will be a sea battle tomorrow', as uttered by you yesterday, was true at the time of utterance.
>
> The intuition that by asserting (2) you have said something true is very strong. In fact, it would be extremely odd to deny at *t*2 that 'There will be a sea battle tomorrow', as uttered by me at *t*1, was true at the time of utterance. (2008: 329)

Granted: *you*—dear reader—might utter such a monstrosity as (2), philosophically inclined as you are likely to be. But (2) is not something you are likely to hear in an ordinary setting. At any rate, *I* for one do not find it extremely odd to suppose that 'There will be a sea battle' was not true at the time of the given utterance. Again, what I instead find extremely odd to deny is that your prediction (that there would be a sea-battle) came true (came to pass, was borne out, was fulfilled, was vindicated).[37]

But let me now change direction. It is not for no reason that the argument against the open future from the validity of Retro-closure has been so popular. (And it is not for no reason that some open futurists have tried—in vain—to preserve it.) The principle *is* an intuitive principle—even if, as I have maintained above, it is not as perfectly obvious as many have seemed to think. We open futurists must give up an intuitive principle; this much is clear. The only thing to say at this stage is the usual appeal to reflective equilibrium—of the overall need to balance, for instance, the metaphysical principles we accept with what "ordinary judgments" we might want to preserve. In the end, I contend that *more is at stake* in the preservation of the open future than is at stake in the preservation of Retro-closure. At stake in the open future is the metaphysical concern that there should not be a realm of fact about the future that outstrips what reality thus far mandates or otherwise determines. And at stake in the preservation of Retro-closure, I contend, is merely the preservation of "what we ordinarily say"—for example, that "you were right". Part of the broader question we must face, if we are open futurists, is *how much of what we say is untrue or false*—and how willing we are to tolerate the result that much of what we say *is* untrue or false. And this is the topic I pursue in the final chapter.

[37] Essentially the same response may be brought to bear on Lowe's (2002: 323) presentation of a similar point.

8
The Assertion Problem

Peter van Inwagen has written, "Well, I suppose I am enough of a Wittgensteinian to think that it is not possible for very much of what we say 'in the midst of life' to be false".[1] Well, I suppose I am not. And my goal in this chapter is to explain how it is that, even though, according to the view defended in this book, much of what we say in the midst of life is false, no one is under any pressure—moral, philosophical, or otherwise—to do anything about it. More generally, my aim in this chapter is to display how the view defended in this book interacts with standard *norms of assertion*. In particular, some philosophers have described a certain sort of problem for open future views such as my own—a problem they have called "the assertion problem". Precisely what the "assertion problem" is meant to be is, I think, unclear—but the basic problem can perhaps be stated as follows: we often assert claims about the future, and we seem rationally warranted in doing so—but the theory of the open future would imply that many such claims are untrue or false; but since... well, something or other, we have a problem for the doctrine of the open future.

As I said: precisely what the "assertion problem" is meant to be is, in my judgment, unclear. My contention is that when we encounter "the assertion problem", what we encounter is really a family of related problems. In this chapter, I try to articulate and respond to this family of problems, and I do so by developing two related themes. First, where we do assert future contingents, and seem rationally warranted in doing so, the open futurist can argue that although, in these cases, we assert what is untrue or false, we nevertheless *communicate* what is true—and that this explains the warrant in question. Second, and relatedly, she can offer *replacement talk*—that is, she can show that, in principle, we could replace our current talk (in which we assert future contingents) with nearby talk (in which we don't), and that we could do so without sacrificing anything of genuine importance.

In this chapter, I rely at various points on a comparison with another theory in metaphysics and philosophy of language that also delivers the result that much of what we say "in the midst of life" is false. (This is the theory at issue in the context of van Inwagen's quote above.) According to certain *eliminativist* views in ontology, there are no such things as, say, chairs and tables and the like—in reality, all

[1] Van Inwagen 2014: 8.

The Open Future: Why Future Contingents are All False. Patrick Todd, Oxford University Press. © Patrick Todd 2021.
DOI: 10.1093/oso/9780192897916.003.0009

we have is *atoms arranged chairwise* or *atoms arranged tablewise*, and so on. (And, thankfully, atoms arranged chairwise are good enough for sitting.) According to Trenton Merricks, for instance, when someone says, in the ordinary course of life, "There are a few chairs in the next room over", what that person says is strictly speaking false: there are no *chairs* in the next room over, for there are no chairs *period.* However, there *are* atoms arranged chairwise in the next room over (or certainly may be)—and this explains why "There are a few chairs in the next room" is intuitively a "better thing to say" than "There are a few unicorns in the next room"—despite the fact that, on Merricks' view, both such claims are false.[2] Needless to say, I do not hereby mean to endorse the relevant kind of ontological eliminativism. I mean only to point out that this kind of eliminativism is attractive (at least to many) on metaphysical grounds—and that this sort of position provides a useful point of contact with what might be said by proponents of open future views like my own.

8.1 The First Problem: Must Open Futurists Change Their Ways?

The first problem I wish to address pertains to a suspicion about the consequences of accepting the theory propounded in this book. The suspicion is that if we accept this theory, we should then have to *reform* in some absurd or unattractive way—in particular, we would have to (in some sense of "have to") refrain from asserting claims we otherwise would have felt perfectly comfortable asserting, or perhaps even attempt to correct others when they assert the relevant claims. In the course of bringing out this sort of problem for open futurists, for instance, Christopher Hughes writes the following:

> Consider the following (three-way) conversation, (CV1):
>
> *A*: It's inevitable that she'll marry him.
> *B*: That's not true.
> *C* (*to B*): But didn't you say she could (still) marry him?
> *B* (*to C*): I did say that, and I still think it's true. But what *A* said is that she'll inevitably marry him, and that's not true.
>
> Here, although *C's* question to *B* is odd, there is nothing odd about *B's* part in the conversation. Now consider the following variant of (CV1), (CV2):
>
> *A*: She'll marry him.
> *B*: That's not true.
> *C* (*to B*): But didn't you say she could (still) marry him?
> *B* (*to C*): I did say that, and I still think it's true. But what *A* said is that she'll marry him, and that's not true.

[2] Merricks 2001.

> In (CV2), I want to say, *C's* question to *B* is perfectly natural, and *B's* part in the conversation is odd. It's perfectly fine to acknowledge that she could still marry him, and also assert (flatly, without any hedging or qualification) that *she will inevitably marry him* is not true. But it's at least odd to acknowledge that she could still marry him, and also assert (flatly, without any hedging or qualification) that *she will marry him* is not true. Certainly, I would not do that, and I cannot recall ever having heard anyone else do it. If someone tells me that she'll marry him, and I believe that it's still possible but not yet inevitable that she'll marry him, then I might naturally say (without any hedging or qualification), "that's not necessarily true", or "that's not inevitable", but I wouldn't naturally say "not true!".[3]

Granted. Neither would I—and I'm an open futurist (at least for the sake of this conversation). I would not, in the ordinary course of life, say "that's not true", in response to "She will marry him", merely on grounds that it is *open* whether she marries him. But what is the significance of this fact? As I hope to explain: not much. Hughes goes on:

> Believers in privilege [a privileged branch, viz., Ockhamists] have no difficulty explaining the (apparent) fact that *B*'s contribution to (CV2) is odd. They can say:
>
> > You are not supposed to assert something flatly (without hedging or qualification) unless you take yourself to know it. [This is the "knowledge norm of assertion".[4]] And in (CV2), *B* shouldn't take himself to know what he flatly asserts. For *B* himself has admitted that there are now-possible histories in which she'll marry him, and should admit that for all he knows, the privileged now-possible history is such a history.
>
> Egalitarians [open futurists, i.e., deniers of a privileged branch], on the other hand, seem ill-placed to explain (or explain away) the (apparent) oddity of *B's* contribution to (CV2). They admit that, in the circumstances, there would be

[3] Hughes 2012: 56–57. For a similar point, see Besson and Hattiangadi 2020: 11. See further Santelli 2020. Let me take this opportunity to make one observation about the "assertion problem" described especially in the latter two texts. In these papers, it is claimed that *we often comfortably assert future contingents*—and the problem for the open future view, such as it is, is meant to follow from this key observation. But this observation is ambiguous between two readings:

(1) We often happily assert what are in fact future contingents
(2) We often happily assert what are future contingents *under that description*

But whereas I am certainly prepared to grant (1), I am less sure about (2)—at least, (2) is significantly less plausible than (1). Are we often perfectly prepared to assert that something will happen, whilst in the same breath acknowledging that it could fail to happen? Perhaps we are, but this certainly isn't clear; at any rate, the issue here seems parallel to the contested issue of whether we are happy to assert "lottery" sentences, e.g., that our ticket will fail to win. Are we perfectly happy to assert that our ticket will fail to win, when we also grant that, objectively, there is perfectly possible (albeit very unlikely) continuation of the present circumstances in which it *does* win? I don't know.

[4] The literature on the knowledge norm is of course enormous; for a start, see Williamson 1996, Benton 2014, and Simion and Kelp 2017.

> nothing untoward in *B's* asserting, without hedging or qualification, that it's open whether she'll marry him. And they think that *it's open whether she'll marry him* implies *it's not true that she'll marry him.* So why should there be anything untoward about *B's* asserting, without hedging or qualification, that it's not true that she'll marry him?[5]

Answer: because there is something untoward about bringing to bear one's obscure—even if perfectly justified—philosophical theory in an ordinary context in which the falsity of such a theory is taken for granted. The theory that *it's* (causally) *open whether she'll marry him* implies *it's not true that she'll marry him* is a philosopher's theory. It is, I contend (Chapter 1), a good such theory—but it remains a philosopher's theory. The problem with *B*'s assertion, in this context, is thus that it reveals *B* to be a philosopher who has forgotten that he is not in the company of fellow philosophers discussing a theory, and who is—recalling themes from Chapter 5—absurdly insistent on inflicting his philosophical theory on his unsuspecting peers. But this says more about *B*, I say, than it says about the reasonability of *B*'s theory. Look at it this way. Suppose *B* said, not only "That's not true", but immediately and without pause added, "but then again, it's *also* not true that she will *not* marry him—presentism being true, and it also being the case that...". What should we make of *B*'s contribution *then*? Well, we should certainly think that it was untoward, surprising, and out of place. But we should think that, I contend, because it would be immediately apparent that *B* is invoking some kind of philosophical theory in a context in which we weren't doing any philosophizing. Indeed, on reflection (in the "philosophy room", as van Inwagen is wont to say), we may even grant that *B*'s theory is an interesting one, and perhaps even, well, *true.* But still. Why can't *B* just, well, *relax*, and talk like a normal person?

Hughes goes on—and it is here that we come to what can appear to be *normative* claims about what open futurists *should* and *should not* do:

> There are also cases of apparent conversational non-untowardness that privilegists [Ockhamists] have no trouble explaining, but egalitarians [open futurists] have trouble explaining (or explaining away). Consider (CV3):
>
> A: She'll marry him.
> B: I doubt that's true, though I grant things could go either way.
> Here *B's* contribution to (CV3) seems perfectly in order. This raises problems for egalitarians. If you think that *p* is incompatible with *q*, you shouldn't say, "I doubt that *p*, though I grant that *q*": the adversative ("though") is inapposite. (Compare:

[5] Hughes 2012: 57.

> "I doubt that's a golden retriever, though I grant that it's a guinea pig".) For example, if you think (as you should), that *it is inevitable that p* is incompatible with *p could still go either way*, you shouldn't say, "I doubt it's inevitable that *p*, though I grant *p* could go either way". And if someone else says to you, "I doubt it's inevitable that *p*, though I grant *p* could go either way", you should respond that that's not the right way to put things. Now egalitarians think that just as *it is inevitable that p* is incompatible with *p could still go either way, it is true that p* is incompatible with *p could go either way*. So they should never say, "I doubt that's true, though I grant things could go either way". And if someone else says that to them, they should respond, "that's not the right way to put things".[6] [Underlined emphasis added]

My response to this complaint is simple: who is Hughes to tell us open futurists how to live our lives?

Slightly more seriously: of course I understand the *thought* behind Hughes' insistence that open futurists should not say, e.g., "I doubt that's true, though I grant things could go either way". And I understand the thought behind the claim that we should respond, in the given scenario, with "that's not the right way to put things". But on this latter point, we need to appreciate precisely what Hughes seems to be suggesting. Hughes seems to be suggesting that we open futurists (the handful of us that there are) should—does he mean *really* should?—be prepared to *correct* people in ordinary life who say things that are inconsistent with the truth of our theory. By way of reminder: this is a philosophical theory. And it is a philosophical theory that is more or less unknown in the public at large. And it is a philosophical theory that is rejected by the majority of even the *philosophers* who are aware of it. And yet: according to Hughes, at least at first blush, we tiny minority of believers in the philosophical theory of the open future have the solemn duty to go around saying, in the relevant cases, to people who don't know about, care about, or accept our theory, "I'm sorry, but that's not the right way to put things". But no. We don't. We have better things to do with our time (viz., pretty much anything) than to attempt to correct people in the way envisaged here by Hughes. And this is true even if our theory is true. Our theory, *even if true*, sadly, is just *not that important*.

It is here that we can make our first comparison with the doctrine of ontological eliminativism. As explained: the eliminativist believes that there are no tables and chairs—really, there are only atoms arranged tablewise, and atoms arranged chairwise, and so on. Now consider. What follows concerning what the eliminativist should and should not assert, and what follows concerning what assertions (from others) the eliminativist should and should not attempt to correct? On the

[6] Hughes 2012: 57–58.

latter issue first: suppose someone suggested the following. Suppose someone suggested that Trenton Merricks—the arch-defender of the given kind of eliminativism—should, given his theory, *correct* people who, in the ordinary course of life, assert, for instance, that there are tables in an adjacent room. "Pardon me", he might say, "but that is not the right way to put things: there are no *tables* in the next room (there being no tables *period*)—but there are certainly atoms arranged tablewise in the next room". It is perfectly obvious that Merricks is under no obligation, given his acceptance of eliminativism, to attempt to correct any such ordinary person speaking in the midst of life. To suggest that this what Merricks *should* do, given that he accepts the relevant theory, is to suggest that Merricks should, *inter alia*, consistently waste his own time, and annoy and confuse a host of innocent bystanders in the process. In short: it is simply false that Merricks should respond in the imagined way *in ordinary life*. And the explanation of this fact is simple: *in ordinary life*, no one takes eliminativism seriously, and further, and crucially, the truth of eliminativism is not the kind of truth which is such that, *if* it were true, it would be deeply important for ordinary people to know about it. (It also seems relevant that, even if it *were* deeply important for ordinary people to know about it, the chances of Merricks successfully convincing ordinary people that it is true are approximately 0.) Thus: even if he believes eliminativism, and even if eliminativism is true, it is not the case that Merricks *should* (attempt to) correct people in the imagined way in ordinary life. And, I suggest, the same is true of himself: it is not the case that Merricks *should* (attempt to) monitor *himself* in ordinary life, making sure, for instance, always to talk in terms of atoms arranged table wise, and never in terms of tables. That would be a terrific waste of mental energy, and more else besides. *It just doesn't matter.*

Plainly, many philosophical views are importantly different, or at least arguably so. There is something (at least potentially) discomfiting about a philosopher who accepts atheism on philosophical grounds, but happily says, in the midst of life, that God exists and cares for us. Similarly, believing in moral responsibility skepticism may give one a *prima facie* or *pro tanto* duty (more on these notions shortly) to attempt to correct people who say, for instance, that a given person is blameworthy for what he has done. (After all, if someone asserts that someone is blameworthy, and one believes that he isn't, failing to correct the record can seem to be seriously unjust.) But it is perfectly obvious that some philosophical views are *not* of this kind—and that eliminativism is one of them. There are no comparable issues, as far as I can see, that arise with respect to "correcting the record" if someone says that there are chairs in the adjacent room, and there aren't—because there are only atoms arranged chairwise in the next room. And the same goes, I say, for open futurists operating in the "midst of life". There is no important issue at stake, for instance, in attempting to correct someone who says (in the relevant instance) "I doubt that's true, though I grant things could go either way", and there is no important good to be achieved if open futurists tried to prevent *themselves* from talking this way in ordinary life. We open futurists are operating in a non-open futurist's world. We are under no obligation to act otherwise.[7]

But perhaps I am misconstruing Hughes' basic point. Perhaps we can leave the relevant "shoulds" out of the picture. Perhaps Hughes' basic point just is that, just as we recognize that "I doubt that's a golden retriever, though I grant that it's a guinea pig" is an inappropriate thing to say (in some sense of "inappropriate"), so the open futurist must maintain that (in the relevant cases) "I doubt that's true, though I grant things could go either way" is inappropriate (in just that sense of "inappropriate")—which it isn't. But this is false—or anyway misleading. The open futurist can clearly admit that, given the operative assumptions of ordinary life, "I doubt that's true, though I grant things could go either way" is perfectly appropriate. As I have emphasized throughout this book, however, the open futurist simply rejects these operative assumptions. My argument for the open future does not proceed from anything like the "inappropriateness" of saying "I doubt that's true, though I grant that things could go either way", but instead seeks to *overcome* this appropriateness through philosophical argument and reflection. Of course, given normal assumptions, this is a perfectly understandable, perfectly appropriate thing to say—but, once we pause to think about the matter more carefully, why accept the assumptions in question?

But suppose we take the relevant "shoulds" seriously. Here we must consider what may seem to be an obvious fallback position for Hughes, and for those sympathetic with his basic point. One might insist that Hughes' basic point should all along be stated, not in terms of an *all things considered* "should", but in terms of a *prima facie* "should", or perhaps in terms of *pro tanto* reasons—reasons that carry the day, as it were, unless they are outweighed by competing considerations. After all, plainly Hughes needn't be committed to the outlandish theory that, no matter what, given what open futurists believe, they should always correct someone who says "I doubt that's true, though I grant that things could go either way". Hughes' point isn't that the open futurist should attempt this correction, even if, as a matter of causal necessity, she'll be promptly shot by the person she corrects. (Similarly, if you'll be shot if you correct the person who says "I doubt that's a golden retriever, though I grant that it's a guinea pig", you shouldn't make that correction—but that is the *all things considered* "should"; you still, perhaps, have

[7] The point I am making here is, I hope, consistent with the claim (ably defended by Bailey and Brenner 2020) that there are indeed contexts in which composition matters; similarly, I am happy to say that there are indeed contexts in which the truth of open futurism would matter. The point I am making, however, is that even if there are contexts in which the truth of open futurism would matter, it doesn't follow that the truth of open futurism would matter so much that we should, in the imagined ways, in ordinary settings, try to prevent ourselves from saying what is strictly untrue or false by the lights of that theory. (I thank an anonymous referee for raising this issue, and Andrew Bailey for helpful discussion.)

pro tanto reason to make that correction, but such reason is simply being outweighed.) Hughes' point, perhaps, is simply that the open futurist has some *pro tanto* reason to make such a correction—and here, Hughes may say, it is simply false that anyone has even such a *pro tanto* reason. But now, I contend, what seemed to be some deep problem for the open futurist disappears. For it is not absurd to maintain, as an open futurist, that one has some small *pro tanto* reason to correct someone who says, "I doubt that's true, though I grant that things could go either way". After all, this small *pro tanto* reason is immediately and vastly outweighed by the fact that such an attempted correction is extraordinarily unlikely to succeed in the ordinary course of life—and the importance of that correction is simply not that great to mandate its attempt despite the odds.

8.2 Another Assertion Problem

However we are to interpret Hughes' argument above, we should see how the points in question can allow us to respond to one way someone might be tempted to construe an "assertion problem" for the open future. Consider an objection as follows:

> Say that a train is scheduled by a reliable provider to arrive at 4:30 PM. On Ockhamism, if there is a small objective chance of the train's *not* arriving at 4:30, it is still overwhelmingly likely that it is nevertheless true that it *will* arrive at 4:30, and so still overwhelmingly likely that if I *assert* that the train will arrive at 4:30, I shall be asserting what is true. Hence, in normal circumstances, it is clear how, given Ockhamism, I could be rationally warranted in asserting that the train will arrive at 4:30. But on the open future, if there is a small objective chance of the train's *not* arriving at 4:30, it is *definitely not true* (and, on the current view, false) that it *will* arrive on time, and so, in view of that small objective chance, if I *assert* that the train will arrive at 4:30, I shall be asserting what is untrue or false. So if I believe the current view, and if I believe that there is a small objective chance of the train's not arriving as scheduled at 4:30, I should not assert any such thing as "Yeah, his train will arrive at 4:30, so you better go pick up him". But of course we say such things all the time, and our doing so seems perfectly appropriate.[8]

As I see it, everything in the above is perfectly in order, except for the second-to-last line—the line which attempts to draw some sort of *normative* conclusion from

[8] This objection has been inspired by (but is only very loosely modeled on) MacFarlane 2014: 230–231.

the truth, or the acceptance, of the doctrine of the open future. The objector maintains that if we accept that there is a small objective chance of the train not arriving at 4:30, and that it is a consequence of this fact that it isn't *true* that the train will arrive at 4:30, then it follows that we should not *assert*, in the normal course of life, that the train will arrive at 4:30. And this is what I mean to deny. I maintain that it is a consequence of the fact that there is a small objective chance of its becoming the case (in n units of time) that ~*p* that it isn't *true* (indeed, is false) that it will become the case (in n units of time) that *p*. But I flatly deny that my acceptance of this thesis implies that, in the ordinary course of life, in the relevant cases, I should try to stop myself from asserting the given instance of the claim that it will be the case that *p*.

Look at it this way. If it is false that there are chairs in the adjacent room *on ordinary grounds*, and one knows this, then, yes, it would be *prima facie* problematic to assert that there are chairs in the adjacent room. But the reason for this is simple. If it is false that there are chairs in the adjacent room on *ordinary* grounds, then anyone attempting to find something for sitting in the next room will be seriously inconvenienced, because in this case there won't even be *atoms arranged chairwise* in the adjacent room. But if it is false that there are chairs in the adjacent room on grounds defended by *Merricks*, it is *not* similarly problematic to assert, in the midst of life, that there are chairs in the adjacent room; for even though this assertion is strictly speaking false, it nevertheless communicates something that is true, viz., that in the adjacent room (well, don't eliminate "rooms" for the minute) there are atoms arranged chairwise, and those are good enough for sitting.

Similarly, if it is false *on ordinary grounds* that the train will arrive at 4:30, and one knows this, then, yes, it would be *prima facie* problematic to assert that the train will at 4:30. But the reason for this is simple. If it is false that the train will arrive at 4:30 *on ordinary grounds*, then the train *won't* arrive at 4:30—in which case anyone counting on its arriving at 4:30 may be seriously inconvenienced. But if it is false that the train will arrive at 4:30 on grounds defended by *me*, it is not similarly problematic to assert, for the usual reasons, in ordinary life, that the train will arrive at 4:30. In other words, in our ordinary, daily practice, there is of course some small risk that we may take when saying, for instance, that someone's train will arrive at 4:30, so you'd better go pick that person up. There is some small risk, of course, that the given person's train may end up delayed, in which case the person going to pick the other person up may be inconvenienced. But we do not, in our daily, ordinary practice, take this as reason enough to hedge or otherwise refrain from saying, e.g., "Yes, his train will arrive at 4:30, so you'd better go pick him up". And my point is that if this small risk isn't reason enough for the non-open futurist to refrain from making the relevant assertions, then it isn't enough for the open futurist to do so either.

8.3 Asserting What Is False, but Communicating What Is True

My response to the above problem relies on a familiar thesis: that an assertion may assert what is false but nevertheless communicate what is true. Further, that an assertion asserts what is false but communicates what is true can explain the *appropriateness* of that assertion. Finally, the idea that the relevant falsehoods can nevertheless convey truths is intimately related to the idea of *replacement talk*—viz., that we could, if we wished, jettison the talk that is false in exchange for talk in terms of the truths those falsehoods typically *convey*. Here it is worth considering a few applications of this point in the present context. Consider first an example concerning ontological eliminativism:

> "There is a chair nearby".
> Falsehood asserted: There is a chair nearby.
> Truth communicated/suggested replacement talk: There are atoms arranged chairwise nearby.

And we can say something similar in the case at hand. (In each case we assume that the relevant event is not a future-necessity.) Consider a few examples:

> "His train will arrive at 4:30".
>
> Falsehood asserted: It will be that his train arrives at 4:30.
>
> Truth communicated/suggested replacement talk: The train is scheduled to arrive at 4:30.
>
> If things go according to plan, the train will arrive at 4:30.
>
> Smith asks me: "Hey, are you going to the conference in Barcelona next month? There are some things we need to discuss". I say: "Yeah, I'll be there, so let's talk then".
>
> Falsehood asserted: It will be that I am at the conference next month.
>
> Truth communicated: I intend/plan to be at the conference next month.
>
> If things go according to plan, I will be at the conference next month.

General observation: this strategy will be available whenever the relevant *will* sentence conveys *plans* or *intentions*. Thesis to be defended: most of the clear, uncontroversial cases of what are perfectly appropriate assertions of future contingents will be cases in which the relevant *will* conveys plans or intentions. (The others will convey a worldly *tendency*; I turn to such cases below.) In this regard, it is useful to consider whether and when we would be comfortable replacing what we *usually* say for certain less idiomatic ways of saying the same thing, but which arguably do not convey plans or intentions. Consider a set of our standard valedictions—that is, our ways of parting ways. Context: Tuesday 5:45 PM, at the end of a long workday. Jack says to his co-workers:

(1) God, it's been a long day. I'll see you guys tomorrow.

But now compare the utterly common and unremarkable (1) with:

(2) God, it's been a long day. It will be that I see you guys tomorrow.

Why is (2) such an odd thing to utter, whereas (1) is completely unremarkable—despite the fact that (1) and (2) are identical in terms of asserted content? Certainly one tempting answer here is the following: in an ordinary context, (1) conveys something like the speaker's plan or intention to leave now but to be back tomorrow, whereas (2) is ambiguous in this regard—it is something you would say only if, strangely, you wanted to be *understood as* making a prediction about tomorrow, and not merely conveying one's *plans* about tomorrow. Thus we can say the following:

"God, it's been a long day. I'll see you guys tomorrow".
Falsehood asserted: It will be that I see you guys tomorrow.
Truth communicated: I plan to see you guys tomorrow.
If things go according to plan, I'll see you guys tomorrow.
"God, it's been a long day. It will be that I see you guys tomorrow".
Falsehood asserted: It will be that I see you guys tomorrow.
Truth communicated: ??

Hence, we have an explanation of the appropriateness of the former and the inappropriateness of the latter.[9]

Let me make one further important point at this stage. Note that in the above I have stated two true sentences that the relevant falsehoods arguably convey: the

[9] One *could* pursue the route of maintaining that (1) does not assert any prediction at all. In general, I think this is the wrong strategy with respect to (1). However, it is worth bringing out the point that not all *will* sentences are predictive. Consider this example (lightly modified) from Copley (2009). You are driving down the highway; note the difference between two billboards:

We will change your oil in Madera.
It will be that we change your oil in Madera.

The first, of course, is readily intelligible; the second is farcical. Clearly, the first is not a *prediction*, but instead an *offer*; it is certainly not "It will be that we change your oil in Madera". The second, however, is something of a threat: it will be that we change your oil in Madera, so you'd better get ready. Other examples of a non-predictive *will* are familiar, e.g., as in the evidential usage, "That will be the postman at the door" (Giannakidou and Mari 2018), and as in the generic, "Oil will float on water" (Kissine 2008). Someone saying that oil will float on water is not necessarily saying that it will be the case, later, that some oil floats on some water. Finally, consider a context in which someone is being summarily dismissed by a commander as follows: "You will leave tomorrow by the first train" (Kissine 2008). Here it makes sense later to ask, "Do you think he'll actually do it? That is, leave by the first train?"—and this is because in making this command, the speaker isn't committed to the *prediction* that he'll leave by the first train.

first is the simple statement of the relevant intention or plan; the latter is the *will* sentence made conditional on its success. Suppose again that we're in a perfectly normal, everyday setting. You ask me whether I'm going to the conference next month; I say, "Yeah, I'll be there". Now, on my account, we can suppose, it is false that it will be that I am at the conference, and false that it will be that I am not at the conference. So what I said was false. Nevertheless, I certainly succeeded in communicating what is true, viz., that I plan to be there. Now, note the following. Holding fixed that I plan to be there, then the relevant *will* claim made conditional on the success of my plan is just plain *true*. In other words, yes, it is false that I will be at the conference (and false that it will be that I am not at the conference). Nevertheless, if things go according to plan, I will be at the conference. In other words: given that I plan to be at the conference, in all the worlds in which my plans are realized, I am at the conference. Thus, it is just plain true that, so long as things go according to plan, I will be at the conference. Note further that the following seems fine:

> There is no fact of the matter regarding whether I'll be at the conference, but so long as things go to plan, I'll be at the conference.

Someone making this assertion is simply asserting that there is no fact of the matter concerning whether her plan will or will not be impeded. I plan to be at the conference, yes. Will I be there in fact? Well, maybe there's no fact of the matter, because there's no fact of the matter whether something will prevent my plans from being realized.

Here I wish to briefly elaborate on my contention that (assuming the relevant plans are in place) I can perfectly well accept the truth of claims such as

(3) If things go according to plan, the train will arrive at 4:30.
(4) If things go according to plan, I will be at the conference next month.

Note: I am not thinking of these sentences as material conditionals. Following Kratzer's general (1991) approach to modals, the view I suggest is that the *if* clause in these statements simply serves to "restrict" the modal at issue—in this case, *will*. Thus, on my view, the above statements do not in fact have future contingents as "consequents" at all. Rather, again, the *if* clause simply restricts the modal. For example, consider

(5) If the lights are on, then Anders must be in his office.

On a Kratzerian analysis, (5) is not analyzed as something of the form ($p \rightarrow$ *Must*: q), where the arrow is material conditional and the *Must* is epistemic necessity. Rather, it is treated as: [(Must: p) q], where the truth conditions of "(Must: p) q" are as follows: "[Must: p] q" is true iff q holds in all the epistemically possible worlds such that *p*. So similarly regarding (3) and (4); iffy-will sentences get the restrictor analysis as well:

"[will: *p*] *q*" is true iff *q* holds in all the available futures such that *p*.

Thus: there is no cause for concern that (3) and (4) are either false or trivially true on my account: they are, I contend, non-trivially true.[10]

8.4 Weakened Readings

One might have the following misgiving about using statements involving plans and intentions as *replacement* talk for our usual talk. Note that, in a usual setting, we sense an important difference between the following replies:

"Hey, are you going to the conference in Barcelona next month? There are some things we need to discuss".

(a) "Yeah, I'll be there".
(b) "I definitely plan/intend to be there".
(c) "So long as things go according to plan, I'll be there".

Someone asserting (b) or (c) would seem to be *worried* about the prospects of his or her making it to the conference. That is, in normal life, you would only say (b) or (c) if there was some especially salient risk to one's plans. Consider:

JANE: Hey, are you going to the conference next month?
SARAH: I definitely plan to be there/So long as things go to plan, I'll be there.
JANE: Oh—is something going on?
SARAH: Well, my mother has an operation scheduled for the week before, and complications could arise, in which case I may have to cancel.
JANE: Ah, got it.

That seems perfectly normal. But now compare:

JANE: Oh—is something going on?
SARAH: Well, you know. Flights are sometimes canceled. Cars sometimes break down. Health emergencies sometimes arise. So I'm just saying that I *plan* to be there; who knows whether I *will be*.

[10] Thanks to Brian Rabern for helpful discussion on this point (and for the *must* example), and thanks to Matt Benton for encouraging me to make these points clearer.

And now Sarah seems overly cautious. (After all, *everyone already knows* that flights are sometimes canceled, and cars sometimes break down, and health emergencies sometimes arise.) Accordingly, one might worry about the suggestion that we could use (b) and (c) as *replacement talk* for what we usually say—viz., (a). The worry, perhaps, is that, if we started saying (b) and (c), we are going to communicate to our interlocutors that our plans are, in some sense, weaker (and more likely to go amiss) than they really are.

The worry is misplaced; indeed, the worry betrays a confusion about what the project of providing "replacement talk" for our talk involving future contingents amounts to in the first place. The worry gets its force from the feeling that if only *some of us* started—on grounds of the open future—saying (b) and (c) in contexts in which we had previously been saying (a)—then those of us who have *not* embarked on this project will be misled when we say (b) and (c) (rather than (a)). And that is certainly true. (They'll think we're more worried about our plans than we really are.) But the idea behind the project of "replacement talk" is not the project of describing how us open futurists should now attempt to talk, given our beliefs. (As I was at pains to emphasize above, it is not the case that anyone who accepts the theory defended in this book should do anything differently at all.) The project is instead the entirely hypothetical (and slightly silly) project of describing what *all of us* could begin saying if *all of us* came to believe the doctrine of the open future (as defended in this book) and wished to avoid saying what is false by the lights of this theory. The project is to show that our talk involving future contingents is *in principle* dispensable, consistently with our living recognizably human lives.

And it is. We could, I take it, simply dispense with all talk involving future contingents, and just talk *directly* in terms of our plans and intentions. Notably, if the question itself is "Will you be at the conference in Barcelona next month?", the response, "I certainly plan to be there" naturally conveys some worry about something that may prevent one's attendance. (Otherwise, why not just say, "Yeah, I'll be there"?[11]) But if the question *just is* from the start "Do you plan to go to the conference?", then the response, "Yes, I certainly plan to be there" does not convey any such worry at all. And my point is that all of us could talk in these terms from the start. And, notably, if we did, some further convention would emerge whereby we indicate which plans we believe are unlikely to be frustrated and which plans we believe are more likely to be frustrated. In other words, "I certainly plan to be there", would have approximately the same force as is now conveyed by "Yes, I'll be there", and "I certainly plan to be there, but I note that . . . " would have approximately the same force as is now conveyed by "I certainly plan to be there".

[11] For interested parties: the relevant Gricean norm here is "assert the stronger".

8.5 Tendencies and the Future

I turn to one further crucial issue. Above we have been focused on cases in which the relevant *will* sentence conveys *plans* or *intentions*. However, it may rightly be felt that there are other cases in which we appropriately assert such sentences, but where nothing about plans or intentions is plausibly conveyed. As a paradigm example, we might simply consider claims about the weather. Suppose John is looking at the weather report for this weekend in Edinburgh. He says, sadly, "That's disappointing, but so typical. Look. It will rain all weekend". Now we can bring out the problem. First, we can note that John's saying "It will rain all weekend" seems appropriate in the circumstances. (At any rate, this much I shall not dispute.) Second, we can suppose that, despite the current weather report, rain this weekend is not yet a future-necessity; there are still causally possible futures in which it fails to rain. Thus, we can suppose that, according to the theory of this book, what John said—viz., that it will rain—is false. So far, so familiar. However, now note that "It will rain all weekend" certainly doesn't convey anything like "The plan is for it to rain all weekend". Thus, it cannot be that John's assertion—though strictly speaking false—is appropriate nevertheless because it communicates some truth about some *plan*.

Granted. Here, however, I contend that it does still convey a truth, and that is a truth about a worldly *tendency*. In short, though strictly speaking John's assertion is false, I contend that it nevertheless conveys the truth that the world is currently *tending towards* rain this weekend. I further contend that it likewise communicates the truth that *if things don't change*, it will rain this weekend. And that much, by my lights, can once more be just plain true. If the weather report is accurate, then the world is now tending towards rain this weekend. Thus, if the world is now tending towards rain in Edinburgh this weekend, it just plain follows that *unless those tendencies change*, there will be rain this weekend in Edinburgh. In other words, in all the worlds where the tendencies remain as they are, there is rain in Edinburgh this weekend. This is the truth that is communicated by John's assertion, and it is because his assertion communicates this truth that it remains an appropriate thing for John to have said, despite its having been false in the circumstances. To wit:

"Look. It will rain all weekend".

Falsehood asserted: It will rain all weekend.

Truth conveyed/suggested replacement talk: The world is tending toward rain all weekend.

If things don't change, it will rain all weekend.

And there we are. Clearly, at this stage, I do not propose to exhaustively consider every last case in which someone appropriately asserts what is plausibly a future contingent. I instead conclude with something of a dilemma. I contend that for any assertion that the theory of this book judges to be false, then either (a) that assertion, though false, communicated a truth regarding either (i) a plan or (ii) a worldly tendency, or (b) that assertion was inappropriate.

It is worthwhile bringing out that, for some of the assertions that the theory of this book implies are false, it is perfectly reasonable to maintain that these assertions are simply inappropriate—in other words, that there is no "assertion problem" arising from the putative observation that the relevant claims are *false* but it is still *appropriate* that we should assert them. Consider a sports fan who asserts, caught up in the moment, that her lowly team will win the game next week. If it isn't even true that the world is now *tending* towards her team's winning next week, then I am happy to say that such an assertion is, well, just plain inappropriate—it is simply this person "going out on a limb", as we sometimes say. However, we do not ordinarily consider someone who, *looking at the weather report*, says "Look. It will rain all weekend" to be similarly "going out on a limb". After all, the latter person conveys a truth about a current worldly tendency, whereas the former person conveys no such truth at all—and is engaged in sheer speculation ungrounded in anything about current reality and the laws governing how it unfolds.

In this context, we might consider whether it could ever be "appropriate" in the relevant sense to assert a claim like (6):

(6) The world isn't tending towards a Giants victory next year, but the Giants will win next year.

This sentence makes perfect sense. (On Ockhamism, many such sentences are even true.) As it happens, however, (6) is exactly the type of utterance the theory of the open future predicts to always be inappropriate. Someone asserting (6) is asserting that the Giants will win next year, not as a matter of a deduction from current trends, but *against* those trends—as if seen in a crystal ball. And it is definitive of the doctrine of the open future that there are *no crystal balls*—more particularly, that there are no truths for such crystal balls in principle to access in the first place. Further, even if there were such truths (viz., truths about what is going to happen, against the current tendencies), presumably even the Ockhamist agrees that *we* do not have access to such truths. Someone asserting (6) is thus always asserting something he could not possibly be in position to know; in any case, the open futurist can happily maintain that assertions in the manner of (6) are always inappropriate. The fact that they are also always false is thus no kind of problem—for the relevant problem arises only if the assertions are false *and* appropriate.

8.6 Interlude on Skepticism

The assertion problem—whatever it is—certainly seems tied up with questions about what we *know* about the future. And it may seem that the current theory must say: hardly anything. Indeed, one might complain that the theory defended in this book results in an unattractive form of skepticism: beyond certain trivialities, we know very little, if anything, about the future.

I am not going to try to fully resolve what proponents of the current position (or open futurists more generally) ought to say in response to this challenge. However, I do want to make one comparison. There are, once more, longstanding parallel issues regarding skepticism and counterfactuals. For instance, Hájek has defended the thesis that "Most Counterfactuals are False" (Hájek ms) on grounds that would seem very similar to those articulated in this book.[12] Now, I am certainly prepared to agree with Hájek that there are no primitive "counterfacts" (Hájek 2020), and thus that what we might call "counterfactual contingents" are all false. However, Hájek adds something to this claim (with which I agree) something with which I am *not* inclined to agree, viz., that "most" (by which Hájek seems to mean nearly all) of the counterfactuals we assert are counterfactual contingents; I am instead prepared to say that a great many are counterfactual necessities. Needless to say, the resolution of this question depends on a critical further question: *what, in general, is causally contingent, and what is causally necessary*? In other words: what is the extent of indeterminism, assuming indeterminism? And this is a question I am certainly not prepared to address in this book.

The important point is the following. My view that future contingents are all false is *not* automatically committed to the extreme position that "most" future-directed claims are false; I am instead inclined toward what I regard to be the moderate position that "much" of our future-directed talk is true—because much of that talk involves the assertion of future-necessities. How much? Well, first of all, I don't know, and second, there is no precise way to say.

But let me at least give a small indication of the position I favor. Consider the claim that I will not be the president of Nigeria in an hour. I think that claim is true. This is because, on reflection, I am prepared to say that it is a future necessity that I shall not become president of Nigeria within the hour. This is because, on reflection, I am prepared to say that it is causally impossible that I should become president of Nigeria within an hour.

There will be certain well-informed readers itching to say, all in a rush, that given certain discoveries regarding quantum mechanics—quantum tunneling this, quantum entanglement that—there is in fact a causally possible future in which

[12] For discussion of counterfactual skepticism, see Lewis 2016, Emery 2017, and Sandgren and Steele 2020.

I am president of Nigeria within the hour. And so, given my view, it is currently false that I will not be president of Nigeria within the hour. It genuinely could happen! It is extremely improbable—but it could happen. All the particles just have to tunnel in the right way, and voila—there I am, the duly-elected president of Nigeria, leader of a bewildered nation.

I am, like any reasonable person, prepared to let the physicists say what the laws of physics permit. However, unlike many reasonable people, I am prepared to insist that, if the laws of physics permit me to become president of Nigeria within the hour, then the laws of physics are not all the laws that there are. For it is not genuinely possible, causally speaking, that I should become president of Nigeria within the hour. If the laws of physics permit this, then there must be other laws that do not, laws that do not simply reduce to those laws—e.g., the laws of psychology. And what is causally possible is what is consistent with the past *and the laws*—by which I mean *all the laws that there are*, whatever they are.

Let me note further the following. I am not saying that, in order to know that something will happen, one has to know that it is a future necessity. My position is not that, in order for Sam to know that he won't be president of Nigeria within the hour, Sam has to know that it is a future necessity that he shall not be president of Nigeria within the hour. (Given all of the difficulties alluded to above regarding the laws of physics, that is a difficult thing to know.) My position is that, in order for Sam to know this, it has to *be* a future necessity that he shall not be president of Nigeria within the hour, regardless of whether this much is known by Sam. Similarly, suppose Rhiannon believes on normal grounds that she will have lunch, as usual, by 3 PM today. My position is not that Rhiannon knows that she'll have lunch by 3 PM only if Rhiannon knows that it is causally determined that she has lunch by 3 PM; my position is that Rhiannon knows that she'll have lunch by 3 PM only if it *is* causally determined that she has lunch by 3 PM.[13] Could this be causally determined? I certainly don't see why not—but who am I to say? At any rate, whether a position of this kind is defensible is a question I am content to leave for another occasion.

8.7 Inshallah

Above I maintained that we *could* refrain from asserting future contingents, and we could do so without substantial practical costs. I wish to conclude by focusing on a particularly dramatic way of bringing out this point, by considering the widespread practice, in many Muslim communities, to preface any statement apparently about the future with *Inshallah*—which is variously translated as

[13] For more on this general distinction, see Alston 1980.

"God willing" or "If God wills it". According to many commentators, this practice is in response to Quaranic command;[14] here is how the practice is explained by a writer for the *Arab News*:

> What does the Arabic "Insha Allah" mean? It's actually a phrase and not one word, but the words are combined into one for easier pronunciation. This phrase is used worldwide by Muslims of all languages, even recognized among many non-Muslims, and it means "God willing". Three words make up this phrase: "In", which means "if"; "Shaa'", which means "will;" and "Allah", meaning "God". Literally, it means "If God wills so".
>
> It's part of every Muslim's daily vocabulary, as we are taught by Islam not to make definitive statements about the future, since only God knows what will happen. This means that if someone asked me to provide him with something, instead of "I will give it to you today" I should say: "I will give it to you today, Insha Allah".[15]

If we take these statements at face value, as I shall, we can make several important observations. First, and most importantly: the "Assertion Problem"—whatever it is—would appear not to arise for the relevant Muslims who observe this command. Apparently, they don't assert claims about the future. (Instead, they would appear to assert claims that are conditional on the content of God's plans.) And if they don't, they don't assert claims that are false by the lights of the current theory.

Consider. On my view, when someone says, "God willing, I'll be at the conference next month", is that a claim that my account predicts is false, on account of the openness of the future? *Prima facie* the answer is no. After all, consider the following exchange:

> KATIE: Hey, are you going to be at the conference in Barcelona next month? There are some things we should discuss.
>
> SARAH: God willing, I'll be there.

[14] "And never say of anything, 'Indeed, I will do that tomorrow', except [when adding], 'If Allah wills'" (Surah 18, verse 23). A very similar exhortation appears in the Christian New Testament (James 4:13–16): "Now listen, you who say, 'Today or tomorrow we will go to this or that city, spend a year there, carry on business and make money'. Why, you do not even know what will happen tomorrow. What is your life? You are a mist that appears for a little while and then vanishes. Instead, you ought to say, 'If it is the Lord's will, we will live and do this or that'. As it is, you boast in your arrogant schemes". This exhortation, as far as I am aware, did not similarly cause a widespread convention to use something like *Inshallah* ("Lord willing...") in the Christian West—or if it did in certain times and places, that practice has fallen out of favor.

[15] "'Insha Allah'—please don't fear this phrase!", *Arab News*. Accessed October 1, 2019 at https://www.arabnews.com/columns/news/715716.

[A MONTH PASSES; SARAH, UNFORTUNATELY, IS PREVENTED FROM ATTENDING BY A SUDDEN CANCELLATION OF HER FLIGHT DUE TO A MECHANICAL BREAKDOWN. ON THE PHONE AFTER THE EVENT:]

KATIE: You know, you were wrong!

SARAH: About what?

KATIE: About being at the conference; you said you were going to be there, and you weren't.

SARAH: Well, I said that *God willing* I would be there—and I guess God wasn't willing; look at what happened with the plane. So I wasn't *wrong*.

Sarah is right. And, it seems, she is right by *anyone's* lights; by the Ockhamist's lights, and my own. She didn't say something false when she said "God willing, I'll be there".

We can thus make the following interim point: if it is a cost for a theory that it makes much of what "we" say false, we can note that, in this context, this cost, whatever it is, will be felt less by those who are already accustomed to the relevant use of *Inshallah*. Once again, however, the real interest of these points is that they show that we *could*—if we wanted to—refrain from asserting future contingents, and that we could do so without substantial practical costs. Above I argued, in effect, that it is not at all important for open futurists to *actually use* such talk. But what *is* important is that there should be such talk; that is, what is important is that there should be a way, in principle, for us to communicate in satisfactory ways, whilst refraining from asserting what is false by the lights of the current theory. And my suggestion is that *Inshallah* points the way towards an account of just this kind.

I will not attempt the difficult project of giving any kind of semantics for "God willing..." or "If God wills it...". I will just note the following. Roughly—and at least as a first pass—when I say that *Inshallah* I'll give you your coat back this afternoon, I'm saying that in all the worlds where my giving you the coat back this afternoon is consistent with God's plans, I give you the coat back this afternoon.[16] And to assert this much is not to assert any future contingent. *Prima facie*, the idea

[16] I said I wouldn't attempt a semantics for "God willing..." or "If God wills it...", but I can't resist noting a few complications anyone wishing to provide such a semantics may wish to take into account. Firstly: the two phrases would not always seem to be equivalent. As a first approximation, "God willing..." is something the speaker would only say if the speaker is in some way positively disposed towards the relevant event; not necessarily so for "If God wills it...". For instance, the following seems fine:

(1) The crops are sadly failing. God willing, there will be rain tomorrow.

But the following seems unintelligible:

(2) We're in real danger of a flood if there's any more rain. God willing, there will be rain tomorrow.

However, (2*) seems fine:

(2*) We're in real danger of a flood if there's any more rain. But if God wills it, there will be rain tomorrow.

here seems closely related to the idea pursued above. Just as, instead of saying, "I'll be at the conference", many already say "Inshallah, I'll be at the conference", so similarly, instead of saying, "I'll be at the conference", we could simply say, "So long as things go according to plan, I'll be at the conference". This would, perhaps, be annoying—but so would talking in terms of atoms arranged chairwise. And that is hardly an objection, in itself, to ontological eliminativism.

8.8 Conclusion

It is a problem for a theory if it makes a great deal of what we say in the midst of life to be *irreplaceably* and *irredeemably* false—that is, false, and such that there is no *nearby* talk we could, if we wished, use to *replace* the talk that is false. But whereas my theory makes a great deal of what we say in the midst of life to be false (assuming the relevant kind of indeterminism), it does not make what we say to be irredeemably false. Further, it is a problem for a theory if it makes a great deal of what we say in the midst of life to be *unintelligibly* false—that is, false, and such that there is no explanation of how it is that we came to use that talk to communicate successfully. But whereas my theory makes much of what we say in the midst of life to be false, it does not make it unintelligibly false. According to the theory defended in this book, much of what we say in the midst of life is

As a first approximation: someone uttering (1) is expressing a hope for rain, whereas someone asserting (2*) is asserting something like God's sovereign power over worldly affairs. A similar contrast can be observed with respect to cases involving the intentions of finite agents—and likewise points to the ambiguity of "If God wills it...". Again, in certain cases, "God willing..." would seem to convey the speaker's intention to bring about the relevant event. Suppose Jack, in the midst of battle, with delusions of future glory, fully intends to die on the battlefield. Jack can say:

(3) I fully intend to die on this battlefield tonight, and so, God willing, I'll die on this battlefield tonight.

But the following seems borderline unintelligible (anyway, a suitably strange scenario would be needed to make it sound fine):

(4) I don't intend to die on this battlefield tonight, but God willing, I'll die on this battlefield tonight.

But note that "If God wills..." can naturally lend itself towards a different reading; unlike (4), the following seems readily intelligible:

(4*) I don't intend to die on this battlefield tonight, but if God wills it, I'll die on this battlefield tonight.

A soldier uttering (4) seems confused—but a soldier uttering (4*) is easily understood. A soldier uttering (4*), like someone uttering (2*), seems to be articulating the commonplace notion that God's will is sufficient for anything God wills—and that God's plans may depart in substantial ways from our plans. (Best to be stoic—as the Stoics maintained—in the face of the possibility that the divine plan departs from one's own.) The overall point: there is a *trivial* reading of "If God wills it...". After all, given the relevant assumptions about God, for any *p*, if God wills *p*, then *p*. "God willing...", however, is less susceptible to such a "trivial" reading, and, in cases involving human subjects and our plans, as some first approximation, it seems to have something more to do with divine *permission* of our plans than with the idea that whatever God wills to happen shall happen. Teasing out these complications, however, is a complicated affair.

(likely) false—but, in the midst of life, I recommended that this fact should simply be ignored. Nevertheless, if you are a *philosopher* working on the open future or related topics, then, when you set down this book, I hope you don't neglect the view I have defended in this book. Since the time of Aristotle, philosophers have taken seriously views on which future contingents fail to be determinately true. I hope we can now take seriously—or more seriously—the position that future contingents are systematically *false*.

References

Adams, Marilyn McCord and Norman Kretzmann. 1969. *William Ockham: Predestination, God's Foreknowledge, and Future Contingents. Translated with an introduction and notes.* New York: Century Philosophy Sourcebooks, Appleton-Century-Crofts.

Alston, William. 1980. "Level-Confusions in Epistemology". *Midwest Studies in Philosophy* 5: 135–150.

Anjum, Rani Lill and Stephen Mumford. 2018. *What Tends to Be: The Philosophy of Dispositional Modality.* Abingdon: Routledge.

Armstrong, D.M. 2004. *Truth and Truthmakers.* Cambridge: Cambridge University Press.

Asay, Jamin and Sam Baron. 2014. "The Hard Road to Presentism". *Pacific Philosophical Quarterly* 95: 314–335.

Azzouni, Jody and Otávio Bueno. 2008. "On What It Takes for There to Be No Fact of the Matter". *Noûs* 42: 753–769.

Bailey, Andrew and Andrew Brenner. 2020. "Why Composition Matters". *Canadian Journal of Philosophy* 50: 934–949.

Barnes, Elizabeth and Ross Cameron. 2009. "The Open Future: Bivalence, Determinism, and Ontology". *Philosophical Studies* 146: 291–309.

Barnes, Elizabeth and Ross Cameron. 2011. "Back to the Open Future". *Philosophical Perspectives* 25: 1–26.

Bartsch, R. 1973. "'Negative transportation' gibt es nicht". *Linguistische Berichte* 27: 1–7.

Belnap, Nuel and Mitchell Green. 1994. "Indeterminism and the Thin Red Line". *Philosophical Perspectives* 8: 365–388.

Belnap, N., M. Perloff, and M. Xu. 2001. *Facing the Future: Agents and Choices in Our Indeterministic World.* Oxford: Oxford University Press.

Bennett, Jonathan. 2003. *A Philosophical Guide to Conditionals.* Oxford: Oxford University Press.

Benton, Matthew. 2012. "Assertion, Knowledge, and Predictions". *Analysis* 72: 102–105.

Benton, Matthew. 2014. "Knowledge Norms". *Internet Encyclopedia of Philosophy.* https://iep.utm.edu/kn-norms/.

Benton, Matthew and John Turri. 2014. "Iffy Predictions and Proper Expectations". *Synthese* 191: 1857–1866.

Berto, Franz and Greg Restall. 2019. "Negation on the Australian Plan". *Journal of Philosophical Logic* 48: 1119–1144.

Besson, Corine and Anandi Hattiangadi. 2020. "Assertion and the Future". In S. Goldberg, ed., *The Oxford Handbook of Assertion.* Oxford: Oxford University Press, 481–504.

Bigelow, John. 1988. *The Reality of Numbers.* Oxford: Oxford University Press.

Bigelow, John. 1996. "Presentism and Properties". In James E. Tomberlin, ed., *Philosophical Perspectives 10: Metaphysics.* Cambridge, MA: Blackwell, 35–52.

Borghini, Andrea and Giuliano Torrengo. 2013. "The Metaphysics of the Thin Red Line". In Fabrice Correia and Andrea Iacona, eds., *Around the Tree: Semantic and Metaphysical Issues Concerning Branching Time and the Open Future.* Dordrecht: Springer, 105–125.

Bourne, Craig. 2006. *A Future for Presentism.* Oxford: Oxford University Press.

Briggs, Rachael and Graeme Forbes. 2012. "The Real Truth about the Unreal Future". *Oxford Studies in Metaphysics* 7: 257–304.

Brogaard, Berit. 2008. "Sea-Battle Semantics". *Philosophical Quarterly* 58: 326–335.

Byerly, T. Ryan. 2014. *The Mechanics of Divine Foreknowledge and Providence: A Time-Ordering Account*. London: Bloomsbury.

Cahn, Steven M. 1967. *Fate, Logic, and Time*. New Haven: Yale University Press.

Caie, Michael. 2012. "Belief and Indeterminacy". *Philosophical Review* 121: 1–54.

Cariani, Fabrizio. 2014. "Attitudes, Deontics and Semantic Neutrality". *Pacific Philosophical Quarterly* 95: 491–511.

Cariani, Fabrizio. 2020. "On Predicting". *Ergo* 7: 339–361.

Cariani, Fabrizio. 2021. *The Modal Future: A Theory of Future-Directed Thought and Talk*. Cambridge: Cambridge University Press.

Cariani, Fabrizio and Paolo Santorio. 2018. "*Will* done Better: Selection Semantics, Future Credence, and Indeterminacy". *Mind* 127: 129–165.

Carter, Jason. "Fatalism and False Futures in *De Interpretatione* 9". Ms. University of St Andrews.

Collins, Chris and Paul Postal. 2014. *Classical NEG Raising: An Essay on the Syntax of Negation*. Cambridge, MA: MIT Press.

Collins, Chris and Paul Postal. 2017. "Disentangling Two Distinct Notions of NEG Raising". Posted on Lingbuzz, https://ling.auf.net/lingbuzz/003595.

Copeland, B. Jack, ed. 1996. *Logic and Reality: Essays on the Legacy of Arthur Prior*. Oxford: Oxford University Press.

Copley, Bridget. 2009. *The Semantics of the Future*. New York: Routledge.

Correia, Fabrice and Sven Rosenkranz. 2018. *Nothing to Come: A Defence of the Growing Block Theory of Time*. Dordrecht: Springer.

Clark, Michael. 1969. "Discourse About the Future". *Royal Institute of Philosophy Lectures* 3: 169–190.

Craig, William Lane. 1987. *The Only Wise God: The Compatibility of Divine Foreknowledge and Human Freedom*. Grand Rapids, MI: Baker.

Craig, William Lane. 2001. "Middle-Knowledge, Truth-Makers, and the 'Grounding Objection'". *Faith and Philosophy* 18: 337–352.

Craig, William Lane and David Hunt. 2013. "Perils of the Open Road". *Faith and Philosophy* 30: 49–71.

Crisp, Thomas. 2007. "Presentism and the Grounding Objection". *Noûs* 41: 90–109.

Dawson, Patrick. 2020. "Hard Presentism". *Synthese*. https://doi.org/10.1007/s11229-020-02580-9.

De Florio, Ciro and Aldo Frigerio. 2019. *Divine Omniscience and Human Free Will: A Logical and Metaphysical Analysis*. Cham: Palgrave Macmillan.

DeRose, Keith. 1994. "Lewis on "Might and "Would" Counterfactual Conditionals". *Canadian Journal of Philosophy* 24: 413–418.

DeRose, Keith. 1999. "Can It Be That It Would Have Been Even Though It Might Not Have Been?" *Philosophical Perspectives* 13: 385–413.

Diekemper, Joseph. 2005. "Presentism and Ontological Symmetry". *Australasian Journal of Philosophy* 83: 223–240.

Dummett, Michael. 1973. *Frege: Philosophy of Language*. London: Duckworth.

Edgington, Dorothy. 1995. "On Conditionals". *Mind* 104: 235–329.

Emery, Nina. 2017. "The Metaphysical Consequences of Counterfactual Skepticism". *Philosophy and Phenomenological Research* 94: 399–432.

Enç, M. 1996. "Tense and Modality". In S. Lappin, ed., *The Handbook of Contemporary Semantic Theory*. Oxford: Blackwell, 345–358.

Fischer, John Martin, ed. 1989. *God, Foreknowledge, and Freedom*. Stanford, CA: Stanford University Press.

Fischer, John Martin. 1994. *The Metaphysics of Free Will: An Essay on Control*. Oxford: Blackwell.

Fischer, John Martin and Patrick Todd, eds. 2015. *Freedom, Fatalism, and Foreknowledge*. Oxford: Oxford University Press.

Flint, Thomas P. 1998. *Divine Providence: The Molinist Account*. Ithaca, NY: Cornell University Press.

Forbes, Graeme. 1996. "Logic, Logical Form, and the Open Future". *Philosophical Perspectives* 10: 73–92.

Freddoso, Alfred. 1988. "Introduction". In Luis de Molina, *On Divine Foreknowledge: Part IV of the Concordia*. Ithaca, NY: Cornell University Press, 1–81.

Gajewski, Jon. 2005. *Neg-Raising: Polarity and Presupposition*. Dissertation. Massachusetts Institute of Technology.

Gajewski, John. 2007. "Neg-Raising and Polarity". *Linguistics and Philosophy* 30: 289–328.

Gaskin, Richard. 1995. *The Sea Battle and the Master Argument*. New York: Walter de Gruyter.

Geach, Peter. 1977. *Providence and Evil*. Cambridge: Cambridge University Press.

Giannakidou, Anastasia and Alda Mari. 2018. "A Unified Analysis of the Future as Epistemic Modality: The View from Greek and Italian". *Natural Language and Linguistic Theory* 36: 85–129.

Goodman, Jeremy. "Consequences of Conditional Excluded Middle". Ms. University of Southern California.

Grandjean, Vincent. 2019. "How Is the Asymmetry between the Open Future and the Fixed Past to Be Characterized?" *Synthese*. https://doi.org/10.1007/s11229-019-02164-2.

Green, Mitchell. 2014. "On Saying What Will Be". In T. Müller, ed., *Nuel Belnap on Indeterminism and Free Action*. Dordrecht: Springer, 147–158.

Green, Mitchell. 2020. "Future Contingents in a Branching Universe". in A. Santelli, ed., *Ockhamism and the Philosophy of Time*. Dordrecht: Springer.

Hájek, Alan. "Most Counterfactuals Are False". Ms. Australian National University.

Hájek, Alan. 2014. "Probabilities of Counterfactuals and Counterfactual Probabilities". *Journal of Applied Logic* 12: 235–251.

Hájek, Alan. 2020. "Contra Counterfactism". *Synthese*. https://doi.org/10.1007/s11229-020-02643-x.

Halpin, John F. 1988. "Indeterminism, Indeterminateness, and Tense Logic". *Journal of Philosophical Logic* 17: 207–219.

Hare, Caspar. 2011. "Obligation and Regret When There Is No Fact of the Matter about What Would Have Happened if You Had not Done What You Did". *Noûs* 45: 190–206.

Hartshorne, Charles. 1941. *Man's Vision of God and the Logic of Theism*. Chicago: Willet, Clark & Company.

Hartshorne, Charles. 1965. "The Meaning of 'Is Going to Be'". *Mind* 74: 46–58.

Hartshorne, Charles. 1970. *Creative Synthesis and Philosophic Method*. La Salle, IL: Open Court.

Hartshorne, Charles and Donald W. Viney. 2001. *Charles Hartshorne's Letters to a Young Philosopher: 1979–1995*, ed. D.W. Viney, *Logos-Sophia*, volume 11. Pittsburg.

Hasker, William. 1985. "Foreknowledge and Necessity". *Faith and Philosophy* 2: 121–156.

Hasker, William. 1989. *God, Time, and Knowledge*. Ithaca, NY: Cornell University Press.

Hasle, Per. 2012. "The Problem of Predestination: As a Prelude to A. N. Prior's Tense Logic". *Synthese* 188: 331–347.

Hawthorne, John. 2005a. "Vagueness and the Mind of God". *Philosophical Studies* 122: 1–25.

Hawthorne, John. 2005b. "Chance and Counterfactuals". *Philosophy and Phenomenological Research* 70: 396–405.

Hess, Elijah. 2017. "The Open Future Square of Opposition: A Defense". *Sophia* 56: 573–587.

Higginbotham, James. 1986. "Linguistic Theory and Davidson's Program in Semantics". In Ernie Lepore, ed., *Truth and Interpretation: Perspectives on the Philosophy of Donald Davidson*. Oxford: Blackwell, 29–48.

Higginbotham, James. 2003. "Conditionals and Compositionality". In John Hawthorne and Dean Zimmerman, eds., *Language and Philosophical Linguistics*, vol. 17. Oxford: Blackwell, 181–194.

Horn, Laurence. 1975. "Neg-Raising Predicates: Toward an Explanation". *Chicago Linguistics Society* 11: 279–294.

Horn, Laurence. 1978. "Remarks on Neg-Raising". In Peter Cole, ed., *Syntax and Semantics 9: Pragmatics*. New York: Academic Press, 129–220.

Horn, Laurence. 1989. *A Natural History of Negation*. Chicago: University of Chicago Press. Reissued, Stanford, CA: CSLI Publications, 2001.

Horn, Laurence. 2015. "On the Contrary: Disjunctive Syllogism and Pragmatic Strengthening". In A. Koslowand A. Buchsbaum, eds., *The Road to Universal Logic*. Dordrecht: Springer, 151–201.

Horn, Laurence R. and Heinrich Wansing. 2017. "Negation". *The Stanford Encyclopedia of Philosophy* (Spring 2017 Edition), ed. Edward N. Zalta. https://plato.stanford.edu/archives/spr2017/entries/negation/.

Hughes, Christopher. 2012. "Openness, Privilege, and Omniscience". *European Journal for Philosophy of Religion* 4: 35–64.

Hughes, Christopher. 2015. "Denying Privilege". *Analytic Philosophy* 56: 210–228.

Hunt, David. 1990. "Middle Knowledge: The 'Foreknowledge Defense'". *International Journal for Philosophy of Religion* 28: 1–24.

Hunt, David. 1993. "Simple Foreknowledge and Divine Providence". *Faith and Philosophy* 10: 394–414.

Iacona, Andrea. "Future Contingents". *Internet Encyclopedia of Philosophy*. https://iep.utm.edu/fut-cont/.

Ingram, David. 2019. *Thisness Presentism: An Essay on Truth, Time, and Ontology*. New York: Routledge.

Kaufmann, Stefan. 2005. "Conditional Truth and Future Reference". *Journal of Semantics* 22: 231–280.

Khoo, Justin. Forthcoming. *The Meaning of If*. Oxford: Oxford University Press.

Kissine, Mikhail. 2008. "Why *Will* Is Not a Modal". *Natural Language Semantics* 16: 129–155.

Knuuttila, Simo. 2014. "Medieval Theories of Future Contingents". *The Stanford Encyclopedia of Philosophy* (Spring 2014 Edition), ed. Edward N. Zalta. http://plato.stanford.edu/archives/spr2014/entries/medieval-futcont/.

Kodaj, Daniel. 2013. "Open Future and Modal Anti-Realism". *Philosophical Studies* 168: 417–438.

Kratzer, Angelika. 1986. "Conditionals". *Chicago Linguistics Society* 22: 1–15.

Kratzer, Angelika. 1991. "Conditionals". In A. von Stechow and D. Wunderlich, eds., *Semantics: An International Handbook of Contemporary Research*. Berlin: De Gruyter, 639–650.

Kvanvig, Jonathan. 1986. *The Possibility of an All-Knowing God*. London: Macmillan.

Leftow, Brian. 1991. *Time and Eternity*. Ithaca, NY: Cornell University Press.

Lewis, David. 1973. *Counterfactuals*. Oxford: Blackwell.

Lewis, Karen S. 2016. "Elusive Counterfactuals". *Noûs* 50: 286–313.

Longenecker, Michael Tze-Sung. 2020. "Future Ontology: Indeterminate Existence or Non-Existence?". *Philosophia* 48: 1493–1500.

Lowe, E.J. 2002. *A Survey of Metaphysics*. Oxford: Oxford University Press.

Lukasiewicz, Jan. 1920. *On Three-Valued Logic*. In *Polish Logic, 1920–1939*, ed. S. McCall. Oxford: Oxford University Press, 1967.

MacFarlane, John. 2003. "Future Contingents and Relative Truth". *Philosophical Quarterly* 53: 321–336.

MacFarlane, John. 2014. *Assessment Sensitivity: Relative Truth and its Applications*. Oxford: Oxford University Press.

Malpass, Alex and Jacek Wawer. 2012. "A Future for the Thin Red Line". *Synthese* 188: 117–142.

Mandelkern, Matthew. 2018. "Talking about Worlds". *Philosophical Perspectives* 32: 298–325 [Corrected version published online 2019, https://doi.org/10.1111/phpe.12112].

Mares, Edwin and Ken Perszyk. 2011. "Molinist Conditionals". In Ken Perszyk, ed., *Molinism: The Contemporary Debate*. Oxford: Oxford University Press, 96–117.

Markosian, Ned. 1995. "The Open Past". *Philosophical Studies* 79: 95–105.

Markosian, Ned. 2013. "The Truth About the Past and the Future". In Fabrice Correia and Andrea Iacona, eds., *Around the Tree: Semantic and Metaphysical Issues Concerning Branching Time and the Open Future*. Dordrecht: Springer, 127–141.

Mehlberg, Henry. 1958. *The Reach of Science*. Toronto: University of Toronto Press.

Merricks, Trenton. 2001. *Objects and Persons*. Oxford: Oxford University Press.

Merricks, Trenton. 2007. *Truth and Ontology*. Oxford: Oxford University Press.

Moruzzi, Sebastiano and Crispin Wright. 2009. "Trumping Assessments and the Aristotelian Future". *Synthese* 166: 309–331.

Moss, Sarah. 2013. "Subjunctive Credences and Semantic Humility". *Philosophy and Phenomenological Research* 87: 251–278.

Müller, T., A. Rumberg, and V. Wagner. 2019. "An Introduction to Real Possibilities, Indeterminism, and Free Will: Three Contingencies of the Debate". *Synthese* 196: 1–10.

Øhrstrøm, Peter. 2019. "A Critical Discussion of Prior's Philosophical and Tense-Logical Analysis of the Ideas of Indeterminism and Human Freedom". *Synthese* 196: 69–85.

Øhrstrøm, Peter and Per Hasle. 2011. "Future Contingents". *The Stanford Encyclopedia of Philosophy* (Summer 2011 Edition), ed. Edward N. Zalta. http://plato.stanford.edu/archives/sum2011/entries/future-contingents/.

Otte, Richard. 1987. "A Defense of Middle Knowledge". *International Journal for Philosophy of Religion* 21: 161–169.

Perloff, Michael and Nuel Belnap. 2012. "Future Contingents and the Battle Tomorrow". *Review of Metaphysics* 64: 581–602.

Pike, Nelson. 1965. "Divine Omniscience and Voluntary Action". *Philosophical Review* 74: 27–46. Reprinted in John Martin Fischer, ed., *God, Foreknowledge, and Freedom*. Stanford, CA: Stanford University Press, 1989, 57–73.

Pike, Nelson. 1970. *God and Timelessness*. New York: Schocken.

Plantinga, Alvin. 1974. *The Nature of Necessity*. Oxford: Oxford University Press.

Plantinga, Alvin. 1985. "Reply to Robert Adams". In James E. Tomberlin and Peter Van Inwagen, eds., *Alvin Plantinga*, Profiles 5. Dordrecht: D. Reidel, 371–382.

Plantinga, Alvin. 1986. "On Ockham's Way Out". *Faith and Philosophy* 3: 235–69. Reprinted in John Martin Fischer, ed., *God, Foreknowledge, and Freedom*. Stanford, CA: Stanford University Press, 1989, 178–215.

Plantinga, Alvin. 1993. "Divine Knowledge". In C. Stephen Evans and Merold Westphal, eds., *Christian Perspectives on Religious Knowledge*. Grand Rapids, MI: Eerdmans, 40–65.

Pooley, Oliver. 2013. "Relativity, the Open Future, and the Passage of Time". *Proceedings of the Aristotelian Society* 113: 321–363.

Prior, A.N. 1957. *Time and Modality*. Oxford: Oxford University Press.

Prior, A.N. 1962. "The Formalities of Omniscience". *Philosophy* 37: 114–129.

Prior, A.N. 1967. *Past, Present, and Future*. Oxford: Oxford University Press.

Prior, A.N. 1976. "It Was to Be". In P. T. Geach and A. J. P. Kenny, eds., *Papers in Logic and Ethics*. Reprinted in John Martin Fischer and Patrick Todd, eds., *Freedom, Fatalism, and Foreknowledge*. Oxford: Oxford University Press, 2015, 317–326.

Pruss, Alexander. 2010. "Probability and the Open Future View". *Faith and Philosophy* 27: 190–196.

Pullum, Geoffrey and Rodney Huddleston. 2002. *The Cambridge Grammar of the English Language*. Cambridge: Cambridge University Press.

Restall, Greg. 2011. "Molinism and the Thin Red Line". In Ken Perszyk, ed., *Molinism: The Contemporary Debate*. Oxford: Oxford University Press, 227–238.

Rhoda, Alan. 2007. "The Philosophical Case for Open Theism". *Philosophia* 35: 301–311.

Rhoda, Alan. 2008. "Generic Open Theism and Some Varieties Thereof". *Religious Studies* 44: 225–234.

Rhoda, Alan. 2009. "Presentism, Truthmakers, and God". *Pacific Philosophical Quarterly* 90: 41–62.

Rhoda, Alan. 2010. "Probability, Truth, and the Openness of the Future: A Reply to Pruss". *Faith and Philosophy* 27: 197–204.

Rhoda, Alan, Gregory A. Boyd, and Thomas G. Belt. 2006. "Open Theism, Omniscience, and the Nature of the Future". *Faith and Philosophy* 23: 432–459.

Robinson, Michael. 2004. "Divine Providence, Simple Foreknowledge, and the 'Metaphysical Principle'". *Religious Studies* 40: 471–483.

Rosenkranz, Sven. 2012. "In Defence of Ockhamism". *Philosophia* 40: 617–631. Reprinted in John Martin Fischer and Patrick Todd, eds., *Freedom, Fatalism, and Foreknowledge*. Oxford: Oxford University Press, 2015, 343–360.

Rumberg, Antje. 2020. "Living in a World of Real Possibilities: Real Possibility, Possible Worlds, and Branching Time". In P. Hasle, D. Jakobsenand , P. Øhrstrøm, eds., *The Metaphysics of Time: Themes from Prior*. Aalborg: Aalborg Universitetsforlag, 343–363.

Sandgren, Alexander and Katie Steele. 2020. "Levelling Counterfactual Skepticism". *Synthese*. https://doi.org/10.1007/s11229-020-02742-9.

Santelli, Alessio. 2020. "Future Contingents, Branching Time and Assertion". *Philosophia*. https://doi.org/10.1007/s11406-020-00235-0.

Schabel, Chris. 2000. *Theology at Paris, 1316–1345: Peter Auriol and the Problem of Divine Foreknowledge and Future Contingents*. Aldershot: Ashgate.

Schoubye, Anders and Brian Rabern. 2017. "Against the Russellian Open Future". *Mind* 126: 1217–1237.

Schulz, Moritz. 2014. "Counterfactuals and Arbitrariness". *Mind* 123: 1021–1055.

Schulz, Moritz. 2017. *Counterfactuals and Probability*. Oxford: Oxford University Press.

Schwarz, Wolfgang. 2018. "Subjunctive Conditional Probability". *Journal of Philosophical Logic* 47: 47–66.
Seymour, Amy. 2014. *Presentism, Propositions, and Persons: A Systematic Case for All-Falsism*. Dissertation. University of Notre Dame.
Sider, T. 2001. *Four-Dimensionalism: An Essay on Persistence and Time*. Oxford: Oxford University Press.
Sider, T. 2006. "Quantifiers and Temporal Ontology". *Mind* 115: 75–97.
Simion, Mona and Christoph Kelp. 2017. "Criticism and Blame in Action and Assertion". *Journal of Philosophy* 114: 76–93.
Stalnaker, R. 1981. "A Defense of Conditional Excluded Middle". In W. Harper, R. Stalnaker and, G. Pearce, eds., *Ifs: Conditionals, Belief, Decision, Chance, and Time*. Dordrecht: D. Reidel, 87–104.
Stefánsson, H. Orri. 2018. "Counterfactual Skepticism and Multidimensional Semantics". *Erkenntnis* 83: 875–898.
Strobach, Niko. 2014. "In Retrospect: Can BST Models be Reinterpreted for What Decision, Speciation Events and Ontogeny Might Have in Common?" In T. Müller, ed., *Nuel Belnap on Indeterminism and Free Action*. Dordrecht: Springer, 257–276.
Stump, Eleonore and Norman Kretzmann. 1981. "Eternity". *Journal of Philosophy* 78: 429–458.
Swinburne, Richard. 2016. *The Coherence of Theism*. Oxford: Oxford University Press. Revised edition; first published 1977.
Tallant, Jonathan and David Ingram. 2020. "A Defence of Lucretian Presentism". *Australasian Journal of Philosophy* 98: 675–690.
Tarski, Alfred. 1935. "The Concept of Truth in Formalized Languages". Reprinted in Alfred Tarski, *Logic, Semantics, Metamathematics*. Oxford: Clarendon Press, 1956, 152–278.
Taylor, Richard. 1957. "The Problem of Future Contingencies". *Philosophical Review* 66: 1–28.
Thomason, Richmond. 1970. "Indeterminist Time and Truth Value Gaps". *Theoria* 36: 264–281.
Todd, Patrick. 2011. "Geachianism". In Jonathan Kvanvig, ed., *Oxford Studies in Philosophy of Religion*, vol. 3. Oxford: Oxford University Press, 222–251.
Todd, Patrick. 2013a. "Soft Facts and Ontological Dependence". *Philosophical Studies* 164: 829–844.
Todd, Patrick. 2013b. "Prepunishment and Explanatory Dependence: A New Argument for Incompatibilism about Foreknowledge and Freedom". *Philosophical Review* 122: 619–639.
Todd, Patrick. 2014. "Against Limited Foreknowledge". *Philosophia* 42: 523–538.
Todd, Patrick. 2016a. "Future Contingents are all False! On Behalf of a Russellian Open Future". *Mind* 125: 775–798.
Todd, Patrick. 2016b. "On Behalf of a Mutable Future". *Synthese* 193: 2077–2095.
Todd, Patrick. 2020. "The Problem of Future Contingents: Scoping out a Solution". *Synthese* 197: 5051–5072.
Todd, Patrick and John Martin Fischer. 2015. "Introduction". In John Martin Fischer and Patrick Todd, eds., *Freedom, Fatalism, and Foreknowledge*. Oxford: Oxford University Press, 1–38.
Todd, Patrick and Brian Rabern. 2021. "Future Contingents and the Logic of Temporal Omniscience". *Noûs* 55: 102–127.
Todd, Patrick and Brian Rabern. Forthcoming. "If Counterfactuals Were Neg-Raisers, Conditional Excluded Middle Wouldn't Be Valid". *Synthese*.

Torre, Stephan. 2011. "The Open Future". *Philosophy Compass* 6: 360–373.
Tuggy, Dale. 2007. "Three Roads to Open Theism". *Faith and Philosophy* 24: 28–51.
Uckelman, Sara L. 2012. "Arthur Prior and Medieval Logic". *Synthese* 188: 349–366.
van Fraassen, Bas C. 1966. "Singular Terms, Truth-Value Gaps, and Free Logic". *Journal of Philosophy* 63: 481–495.
van Inwagen, Peter. 2008. "What Does an Omniscient Being Know About the Future". In Jonathan Kvanvig, ed., *Oxford Studies in Philosophy of Religion*, vol. 1. Oxford: Oxford University Press, 216–230.
van Inwagen, Peter. 2014. *Existence*. Cambridge: Cambridge University Press.
Viney, Donald W. 1989. "God Only Knows? Hartshorne and the Mechanics of Omniscience". In Robert Kane and Stephen H. Phillips, eds., *Hartshorne: Process Philosophy and Theology*. Albany: State University of New York Press, 71–90.
Viney, Donald. 2018. "Process Theism". *The Stanford Encyclopedia of Philosophy* (Summer 2018 Edition), ed. Edward N. Zalta. https://plato.stanford.edu/archives/sum2018/entries/process-theism/.
von Fintel, K. and Iatridou, S. 2002. "If and When If-Clauses Can Restrict Quantifiers". http://web.mit.edu/fintel/www/lpw.mich.pdf.
Wawer, Jacek. 2014. "The Truth about the Future". *Erkenntnis* 79: 365–401.
Wawer, Jacek. 2018. "Some Problems with the Russellian Open Future". *Acta Analytica* 33: 413–425.
Wawer, Jacek and Alex Malpass. 2020. "Back to the Actual Future". *Synthese* 197: 2193–2213.
Whitehead, A.N. 1929. *Process and Reality: An Essay in Cosmology*. Corrected Edition [1978] ed. David Ray Griffin and Donald W. Sherburne. New York: Free Press.
Williams, J.R.G. "Aristotelian Indeterminacy and the Open Future". Ms. Leeds University.
Williams, J.R.G. 2010. "Defending Conditional Excluded Middle". *Noûs* 44: 650–668.
Williams, J.R.G. 2012. "Counterfactual Triviality: A Lewis-Impossibility Argument for Counterfactuals". *Philosophy and Phenomenological Research* 85: 648–670.
Williams, J.R.G. 2014. "Nonclassical Minds and Indeterminate Survival". *Philosophical Review* 123: 379–428.
Williamson, Timothy. 1988. "Bivalence and Subjunctive Conditionals". *Synthese* 75: 405–421.
Williamson, Timothy. 1996. "Knowing and Asserting". *Philosophical Review* 105: 489–523.
Williamson, Timothy and Paal Antonsen. 2010. "Modality & Other Matters: An Interview with Timothy Williamson". *Perspectives: International Postgraduate Journal of Philosophy* 3: 16–29.
Wilson, Jessica. 2016. "Are There Indeterminate States of Affairs? Yes". In Elizabeth Barnes, ed., *Current Controversies in Metaphysics*. New York: Routledge, 105–119.
Woodward, Richard. 2012. "Fictionalism and Incompleteness". *Noûs* 46: 781–790.
Yalcin, Seth. 2010. "Probability Operators". *Philosophy Compass* 5: 916–937.
Zemach, Eddy and David Widerker. 1988. "Facts, Freedom, and Foreknowledge". *Religious Studies* 23: 19–28. Reprinted in John Martin Fischer, ed., *God, Foreknowledge, and Freedom*. Stanford, CA: Stanford University Press, 1989, 111–122.

Index